SpringerBriefs in Physics

SpringerBriefs in Physics are a series of slim high-quality publications encompassing the entire spectrum of physics. Manuscripts for SpringerBriefs in Physics will be evaluated by Springer and by members of the Editorial Board. Proposals and other communication should be sent to your Publishing Editors at Springer.

Featuring compact volumes of 50 to 125 pages (approximately 20,000–45,000 words), Briefs are shorter than a conventional book but longer than a journal article. Thus, Briefs serve as timely, concise tools for students, researchers, and professionals.

Typical texts for publication might include:

- A snapshot review of the current state of a hot or emerging field
- A concise introduction to core concepts that students must understand in order to make independent contributions
- An extended research report giving more details and discussion than is possible in a conventional journal article
- A manual describing underlying principles and best practices for an experimental technique
- An essay exploring new ideas within physics, related philosophical issues, or broader topics such as science and society

Briefs allow authors to present their ideas and readers to absorb them with minimal time investment. Briefs will be published as part of Springer's eBook collection, with millions of users worldwide. In addition, they will be available, just like other books, for individual print and electronic purchase. Briefs are characterized by fast, global electronic dissemination, straightforward publishing agreements, easy-to-use manuscript preparation and formatting guidelines, and expedited production schedules. We aim for publication 8–12 weeks after acceptance.

Folkert Kuipers

Stochastic Mechanics

The Unification of Quantum Mechanics with Brownian Motion

 Springer

Folkert Kuipers
INFN, Sezione di Napoli
Complesso Universitario di Monte
Sant'Angelo
Naples, Italy

ISSN 2191-5423   ISSN 2191-5431 (electronic)
SpringerBriefs in Physics
ISBN 978-3-031-31447-6   ISBN 978-3-031-31448-3 (eBook)
https://doi.org/10.1007/978-3-031-31448-3

This Springer imprint is published by the registered company Springer Nature Switzerland AG
The registered company address is: Gewerbestrasse 11, 6330 Cham, Switzerland

Preface

This monograph is intended as a modern introduction to the theory of stochastic mechanics and its extensions to manifolds using stochastic differential geometry.

Stochastic mechanics is a theory that holds great promise in resolving the mathematical and interpretational issues encountered in the canonical and path integral formulations of quantum theories. It provides an equivalent formulation of quantum theories, but substantiates it with a mathematically rigorous stochastic interpretation by means of a stochastic quantization prescription.

The book builds on recent developments in this theory, and shows that quantum mechanics can be unified with the theory of Brownian motion in a single mathematical framework. Moreover, it discusses the extension of the theory to curved spacetime using second order geometry, and the induced Itô deformations of the spacetime symmetries.

The book is self-contained and provides an extensive review of stochastic mechanics of the single spinless particle. The book builds up the theory on a step-by-step basis. It starts, in Chap. 2, with a review of the classical particle subjected to scalar and vector potentials. In Chap. 3, the theory is extended to the study of a Brownian motion in any potential, by the introduction of Gaussian noise. In Chap. 4, the Gaussian noise is complexified. The result is a complex diffusion theory that contains both Brownian motion and quantum mechanics as a special limit. In Chap. 5, the theory is extended to relativistic diffusion theories. In Chap. 6, the theory is further generalized to the context of pseudo-Riemannian geometry. Finally, in Chap. 7, some interpretational aspects of the stochastic theory are discussed in more detail. The appendices concisely review relevant notions from probability theory, stochastic processes, stochastic calculus, stochastic differential geometry and stochastic variational calculus.

The book is aimed at graduate students and researchers in theoretical physics and applied mathematics with an interest in the foundations of quantum theory and Brownian motion. The book can be used as reference material for courses on and further research in stochastic mechanics, stochastic quantization, diffusion theories on curved spacetimes and quantum gravity.

Naples, Italy   Folkert Kuipers
February 2023

Acknowledgements I am grateful to Michele Arzano, Can Gökler, Qiao Huang and Zacharias Roupas for valuable discussions. This research was carried out in the frame of Programme STAR Plus, financially supported by UniNA and Compagnia di San Paolo.

Contents

About the Author

Dr. Folkert Kuipers is a post-doctoral researcher in quantum gravity at the Istituto Nazionale di Fisica Nucleare (INFN) in Naples, Italy. He holds B.Sc. degrees in Mathematics, Physics and Astronomy (Utrecht University, 2015), M.Sc. degrees in Theoretical Physics and Applied Mathematics (Utrecht University, 2018) and a Ph.D. degree in Theoretical and Mathematical Physics (University of Sussex, 2022).

His research interests range over many aspects of quantum theories on curved spacetimes and quantum gravity. Within these fields, he has contributed to research on effective field theories of quantum gravity. In addition, he made various important contributions to the study of stochastic mechanics and its extensions to curved spacetimes using second order geometry.

For his proposal to apply stochastic differential geometry to the study of quantum gravity, he has been awarded a Humboldt fellowship, which will be carried out at the LMU in Munich.

Chapter 1
Introduction

Abstract This chapter provides an introduction into the theory of stochastic mechanics. It gives a brief historical perspective, and discusses in a qualitative way the main features of stochastic mechanics and its relations to closely related frameworks.

The quantum theory that has been developed since the early 20th century has greatly improved our understanding of fundamental physics. It lies at the heart of all our understanding of both particle physics and condensed matter theory. Moreover, it forms the basis of a large fraction of the technological advancement of the last century. However, despite all these advances, several fundamental aspects of quantum theory are not yet understood. In this regard, the most pressing question is perhaps how to reconcile quantum theory with the theory of gravity that, in its current form, also stems from the early 20th century.

It is widely known that there exist various equivalent formulations of quantum mechanics. Perhaps the best known is Dirac's canonical formalism, which incorporates both Heisenberg's matrix mechanics and Schrödinger's complex diffusion theory. A second widely used approach is the path integral formulation that was later developed by Feynman. It is less known, however, that there exists another formulation of quantum mechanics, which was pioneered by Edward Nelson and is better known as stochastic mechanics. This research program aims to resolve the two major issues that are left by the other two formulations of quantum mechanics.

The first issue is that the canonical formulation of quantum mechanics leaves a rather unsatisfactory picture of physical reality. An orthodox interpretation of the canonical formulation forces one to give up on the notion of a classical configuration space. This classical configuration space only appears, if the probability interpretation is imposed. However, this probability interpretation is ad hoc, as it is not properly defined within modern probability theory. For this reason the foundations of quantum mechanics have been the subject of many debates.

Feynman's path integral formulation, on the other hand, provides a rather intuitive picture of quantum mechanics in which the classical configuration space is retained. However, the path integral formulation has the major drawback that the path integral is a heuristic object and that a precise mathematical definition that applies to all physical theories is still absent.

F. Kuipers, *Stochastic Mechanics*, SpringerBriefs in Physics,
https://doi.org/10.1007/978-3-031-31448-3_1

Stochastic mechanics tries to resolve these two issues by interpreting the paths in Feynman's path integral as the sample paths of a stochastic process. If this process exists, Feynman's path integral is well defined as an Itô integral and can be studied using the tools from stochastic analysis. Moreover, it provides a physical picture in which the quantum fluctuations are similar to the statistical fluctuations encountered in the theory of Brownian motion. Stochastic mechanics thus provides a natural interpretation of quantum mechanics in which the classical configuration space can be retained and the probability interpretation is well defined.

The main challenge faced by stochastic mechanics is to prove the equivalence between stochastic mechanics and quantum theories. Although this has not yet been achieved for all quantum theories, this equivalence has been established for a single spinless particle on a smooth manifold subjected to any potential, and there are strong indications that this equivalence can be generalized to any other quantum theory.

In this book, we review the stochastic theory of a single spinless particle in detail. More precisely, we generalize the stochastic quantization procedure, that was originally formulated by Nelson, and apply this procedure to a single (non-)relativistic spinless particle moving on a pseudo-Riemannian manifold subjected to a vector and scalar potential. As we use a generalized stochastic quantization procedure, we obtain a class of stochastic processes. In a certain limit, this class reduces to the Wiener process that describes Brownian motion, while in another limit it reduces to processes that are similar to the ones studied in stochastic mechanics and describe quantum mechanics.

1.1 A Brief History

The history of Brownian motion can be traced back until at least ca. 60 B.C., when Lucretius hypothesized in his book 'De Rerum Natura' that the existence of atoms should, due to their collisions with larger particles, induce a jiggly motion for the larger particles. It would take until the 18th century, however, that, thanks to developments in microscopy, such a phenomenon was observed in nature by many scientists who studied particles suspended in a liquid [1].

Knowledge about the existence of this phenomenon would become more widespread after Brown published his observation of jiggly motion of pollen suspended in water through a microscope [2]. Due to the increased interest in this phenomenon at the time, this jiggly motion became known as Brownian motion. Since then Brownian motion has become a topic of considerable interest in many branches of science.

In the early 20th century, interest for the phenomenon arose in the physics community, due to the mathematical models developed by Einstein and Smoluchowski [3, 4]. These models were put forward as a way to confirm the existence of atoms and molecules introduced in statistical physics. Soon after, their models were confirmed experimentally by Perrin [5], who thus proved the atomic nature of matter. Later, the models developed by Einstein and Smoluchowski were improved by various other physicists, most notably by Ornstein and Uhlenbeck [6], who built on the theory of stochastic differential equations developed by Langevin [7].

The first mathematically rigorous description of Brownian motion was developed in the framework of functional analysis by Wiener [8]. Due to this work, the stochastic process describing Brownian motion is also known as the Wiener process. Further progress in the mathematical study of Brownian motion was made after the formulation of the axioms of probability theory by Kolmogorov [9]. This allowed to rigorously describe Brownian motion in the framework of stochastic analysis [10], which, in turn, allowed for generalizations to much larger classes of stochastic processes [11], and for the development of stochastic calculus by Itô [12], Fisk [13] and Stratonovich [14]. This calculus provides a rigorous foundation and generalization of the stochastic calculus that was developed earlier by Langevin.

Quantum mechanics arose in the early 20th century, starting from Planck's explanation of the black body spectrum [15] and Einstein's explanation of the photoelectric effect [16]. Both these works served to explain observed phenomena that could not be reconciled with classical physics and quickly gave rise to a radically new theory known as quantum mechanics. Many of the elementary building blocks of quantum mechanics were developed in the early 20th century by Born, Dirac, Heisenberg, Jordan, Von Neumann, Pauli, Schrödinger, Weyl, Wigner and others. This led to two formulations of quantum theory: one based on Schrödinger's complex diffusion equation and one based on Heisenberg's matrix mechanics. These two approaches were later synthesized by Dirac [17], which led to the Dirac-Von Neumann axioms [18]. These axioms form, together with various other postulates, the mathematical basis of quantum mechanics of non-relativistic particles.

After these initial developments, quantum mechanics was extended to quantum field theory, which provided a quantum description of both relativistic and non-relativistic classical field theories. This led to the formulation of the Wightman axioms of quantum field theory [19]. Around the same time, Feynman developed a new formalism of quantum mechanics based on path integrals [20], which was shown to be mathematically equivalent to the canonical formulation of quantum mechanics that was developed earlier by Dirac. This reformulation by Feynman provided a first rigorous connection between the theories of Brownian motion and quantum mechanics, due to its similarities with the Wiener integral that governs the dynamics of a Brownian motion. A more direct connection between quantum mechanics and stochastic theories, such as Brownian motion, was later established by Nelson [1, 21, 22], who introduced another equivalent reformulation of quantum mechanics, better known as stochastic mechanics.

1.2 Diffusion Theories

Given a configuration space $\mathcal{M}$ and an interval $\mathcal{T} \subseteq \mathbb{R}$, a complex diffusion equation is an equation of the form

$$\alpha \frac{\partial}{\partial t} \Psi(x, t) = H(x, \hat{p}, t) \, \Psi(x, t), \tag{1.1}$$

where $\alpha \in \mathbb{C}$, $t \in \mathcal{T}$, $x \in \mathcal{M}$. Moreover, Ψ is a bounded complex valued function and H is a Hamiltonian or diffusion operator. H is second order in $\hat{p}$ and $\hat{p}$ is an operator acting on the space of complex functions on $\mathcal{M}$.

The typical example is the setting where $\mathcal{M} = \mathbb{R}^d$, Ψ is (square-)integrable and $\hat{p} = \pm \alpha \, \partial_x$ is a differential operator, such that H is of the form

$$H = \frac{\alpha^2}{2} \delta^{ij} \left(\frac{\partial}{\partial x^i} + A_i(x, t) \right) \left(\frac{\partial}{\partial x^j} + A_j(x, t) \right) + \mathfrak{U}(x, t), \tag{1.2}$$

where $\mathfrak{U}$ is a scalar field and A is a covector field.

If $\alpha \in (0, \infty)$, the diffusion equation is called the heat equation. The heat equation presents a special limit within all diffusion equations, since it reduces to a real diffusion equation for real valued functions Ψ. A characteristic feature of real diffusion equations is that their solutions are dissipative, meaning that solutions Ψ spread out in the space $\mathcal{M}$, when the time is evolved. This dissipation introduces an asymmetry under time reversal in the theory.

If $\alpha \in \mathrm{i} \times (0, \infty)$, the diffusion equation is called the Schrödinger equation,[1] which is encountered in quantum mechanics. The Schrödinger equation also presents a special limit within all diffusion equations, as it is a conservative diffusion equation [22], which means that the time evolution operator is unitary. As a consequence, the dissipative behavior, that is characteristic to real diffusion, is completely absent in the solutions of the Schrödinger equation. Therefore, the solutions of the Schrödinger equation are symmetric under time reversal.

The heat equation is intimately related to the theory of Brownian motion, because of the Kolmogorov equations [23] and the Feynman-Kac theorem [24]. The Kolmogorov equations describe how the probability density of a continuous-time Markov process evolves over time. There are two such equations. The Kolmogorov forward equations describe the evolution of the probability that a process is in a certain state, given the probability density at an earlier time. The Kolmogorov backward equations describe the evolution of the probability that a process is in a certain state, given the probability density at a later time.

---

[1] The Schrödinger equation is sometimes erroneously referred to as a wave equation, due to its wave-like solutions. We emphasize, therefore, that, by definition, a diffusion equation contains a first order time derivative, while a wave equation contains a second order time derivative, i.e., a wave equation is of the form $\alpha^2 \partial_t^2 \Psi = H \, \Psi$.

The Feynman-Kac theorem states that, given the diffusion equation (1.1) with $\alpha \in (-\infty, 0)$, $\mathcal{T} = [0, T]$, $\Psi : \mathbb{R} \times \mathbb{R}^d \to \mathbb{R}$ and H as given in Eq. (1.2), there exists an $\mathbb{R}^d$-valued stochastic process $X = C + M$, where M is a Wiener process and C a finite variation process, such that

$$\Psi(x, t) = \mathbb{E}\left[\Psi(X_T, T) \exp\left(-\int_t^T \mathfrak{U}(X_s, s)\, ds\right) \,\Big|\, X_t = x\right] \qquad (1.3)$$

for all $t \in \mathcal{T}$. Conversely, given an $\mathbb{R}^d$-valued stochastic process $X = C + M$, where M is a Wiener process and C a finite variation process, there exists a probability measure, such that (1.3) satisfies the diffusion equation (1.1) with $\alpha \in (-\infty, 0)$, $\mathcal{T} = [0, T]$, $\Psi : \mathbb{R} \times \mathbb{R}^d \to \mathbb{R}$ and H as given in Eq. (1.2). We note that a similar result follows for $\alpha \in (0, \infty)$ by considering a stochastic process that evolves backward in time.

The Kolomogorov backward equations can be derived from the Feynman-Kac theorem by noting that the probability that the process X will evolve to the state X_T, given that at time t it is in state X_t, is given by

$$\rho(X_T, T; X_t, t) = \mathbb{E}\left[\Psi(X_T, T) \exp\left(-\int_t^T \mathfrak{U}(X_s, s)\, ds\right) \,\Big|\, X_t\right]. \qquad (1.4)$$

Therefore, the time evolution of the probability density for a Brownian motion is governed by the heat equation. Conversely, for any normalized solution Ψ of the heat equation, one can construct an $\mathbb{R}^d$-valued stochastic process $X = C + M$, where C is a finite variation process and M is a Wiener process, such that Ψ describes the probability density of X.

1.3 The Wick Rotation

The similarities between the Schrödinger equation and the heat equation have been exploited in many ways in quantum physics. At the heart of any such exploitation lies the Wick rotation [25], which can be regarded as a combination of two different transformations:

- the Wick rotation maps a theory on a pseudo-Riemannian manifold onto a theory on a Riemannian manifold;
- the Wick rotation maps a quantum theory onto a statistical theory by mapping the imaginary diffusion constant onto a real diffusion constant.

The Wick rotation thus allows to study quantum theories on pseudo-Riemannian manifolds by mapping them onto statistical theories on Riemannian manifolds. These statistical theories can be studied using the tools from stochastic analysis and the results can be translated back to the original theory by inverting the Wick rotation.

The Wick rotation plays a crucial role in Feynman's reformulation of quantum theories using path integrals. The path integral is an object of the form

$$W(X_f, t_f; X_0, t_0) = \int e^{\frac{i}{\hbar} \int_{t_0}^{t_f} L(X, \dot{X}, t)\, dt}\, DX \qquad (1.5)$$

that integrates over all possible paths that the system X can take in evolving from state $X(t_0) = X_0$ to the state $X(t_f) = X_f$. As these paths are weighted with their probability $e^{\frac{i}{\hbar} S(X)}$, this integral represents the probability amplitude for the system X of evolving from state $X(t_0) = X_0$ to a state $X(t_f) = X_f$.

The path integral (1.5) is a heuristic object, which can be given a precise meaning by mapping it to other functional integrals or to stochastic integrals that have been given a rigorous definition. For the general case this has not yet been established cf. e.g. Ref. [26] for a review. However, if one applies a Wick rotation to (1.5), such that the probability of the sample paths is given by $e^{-\frac{1}{\hbar} S_{\text{Eucl.}}}$, the path integral becomes a Wiener integral, which has a well defined meaning as a functional integral and as an Itô integral. This observation inspired Kac in formulating and proving the Feynman-Kac theorem [24].

The path integral method and the Wick rotation have led to major progress in quantum field theory. Moreover, the Feynman-Kac theorem allows to define Euclidean quantum field theories in mathematically rigorous way [27–31]. Therefore, the path integral approach and the Wick rotation also serve as important tools in constructive quantum field theory, which aims to axiomatize quantum field theory in such a way that all quantum field theories are mathematically well defined.

The similarities between statistical theories and quantum theories have also been exploited in the framework of stochastic quantization developed by Parisi and Wu [32, 33]. In this framework, Euclidean quantum theories are studied by introducing an additional fictious time parameter τ. One then imposes that the theory evolves as a standard Brownian motion with respect to this fictious parameter and studies the equilibrium limit $\tau \to \infty$. In this equilibrium limit, the statistical averages of the extended field theory become identical to the vacuum expectation values of the original Euclidean quantum theory. This allows to employ the methods from statistical physics to study Euclidean quantum field theories.

We emphasize that, although there are various similarities, the stochastic quantization method as proposed by Parisi and Wu is different from the stochastic quantization procedure that is employed in stochastic mechanics and was developed by Nelson and Yasue. Stochastic quantization in the sense of Parisi and Wu will not be studied in this work, whereas the stochastic quantization procedure, as developed by Nelson and Yasue, is the central object of study in this book. Therefore, in the remainder of this book, stochastic quantization will solely refer to the quantization procedure employed in stochastic mechanics, which will be reviewed in the next section.

1.4 Stochastic Mechanics

Stochastic analysis plays a central role in the study of Euclidean quantum theories, which has led to major progress in both rigorous mathematical studies of quantum theories and in numerical simulation of quantum theories. The obvious downside of this use of stochastic tools in the study of quantum theories is that they are applied directly to Euclidean quantum theories, while their application to Lorentzian quantum theories remains indirect, as it relies on the Wick rotation.

Stochastic mechanics is a theoretical framework that aims to apply stochastic analysis to all quantum theories in a direct manner by explicitly constructing stochastic processes that describe the quantum systems. If such a construction can be performed, this has two major advantages:

- it provides a natural interpretation of quantum mechanics as a stochastic theory;
- it allows to apply the stochastic tools that have been very successful in the study of Euclidean quantum theories directly to Lorentzian quantum theories.

Since Euclidean quantum theories of spinless particles are in essence stochastic theories, the major challenge of stochastic mechanics is to construct processes for Lorentzian quantum theories and for theories with spin. In this book, we will focus on the extension to Lorentzian theories of spinless particles. In order to construct stochastic processes that describe such theories, stochastic mechanics must overcome the two issues that are usually resolved by the Wick rotation and were discussed in previous section.

We stress that this provides a major challenge, already for non-relativistic theories, since a straightforward generalization of the Feyman-Kac theorem, involving a single real Wiener process, to complex diffusion equations such as the Schrödinger equation does not exist. More precisely, it has been shown that the complex measure, which is necessary to construct such an equivalence will have an infinite total variation [26, 34, 35]. Therefore, a successful theory of stochastic mechanics must involve a generalization of the standard Wiener process.

In early formulations of stochastic mechanics, this generalization is achieved by considering linear combinations of a future directed and a past directed Wiener process [1, 21, 22, 36]. In more recent formulations [37–39], these two processes have been reformulated as a single complex process. Such processes avoid the issues related to the infinite variation of the complex measure [40]. In this book, we discuss a further generalization of this complex theory suggested in Ref. [41], which allows to describe both stochastic processes associated to quantum mechanics and the ordinary Wiener process in a single framework.

The first theory of stochastic mechanics was developed by Fényes [42], who constructed a stochastic process for which the diffusion equation is given by the Schrödinger equation. A few years later, a similar derivation was developed independently by Kershaw [43] and by Nelson [21]. These derivations of the Schrödinger equation in stochastic mechanics share strong resemblance with the hydrodynamical theory of quantum mechanics that was developed earlier by Madelung [44, 45].

However, in stochastic mechanics the Madelung equations are derived from an underlying theory of stochastic processes.

The initial works by Fényes and Nelson consider a stochastic particle subjected to a stochastic version of Newton's second law, but, soon after, it was realized that this theory could be reformulated in a Lagrangian or Hamiltonian formulation using stochastic variational calculus [22, 46–52]. This reformulation allowed to axiomatize stochastic mechanics by means of a stochastic quantization condition. This condition plays a similar role as the canonical quantization condition and can be formulated as follows:

- the trajectory of a quantum particle can be described by a future directed stochastic process $X(t) = C_+(t) + M_+(t)$;
- the time reversed process is well defined and can be described by a past directed stochastic process $X(t) = C_-(t) + M_-(t)$;
- the processes $C_\pm(t)$ represent a drift that minimizes a stochastic action;
- the processes $M_\pm(t)$ are Wiener processes, i.e. they represent a Gaussian noise with covariance matrix $\mathbb{E}\left[M^i_\pm(t) \, M^j_\pm(t) \mid M_\pm(t_0) \right] = \frac{\hbar}{m} \, \delta^{ij} \, |t - t_0|$.

The theory of stochastic mechanics, as formulated by Nelson, only applies to a single non-relativistic massive spinless particle subjected to a scalar potential and charged under a vector potential in flat spaces $\mathbb{R}^d$ of arbitrary dimension $d \in \mathbb{N}$. Nevertheless, this theory already allows to give a mathematically consistent description of elementary quantum theories such as the free particle and the harmonic oscillator in terms of stochastic processes [22, 36]. Moreover, this stochastic theory provides an interpretation of the single and double slit experiments [22]. In this stochastic interpretation of the double slit experiment, the quantum particle passes only one slit. Nonetheless, the stochastic theory correctly produces an interference pattern, as this is the unique solution of the unitary stochastic process that is studied in stochastic mechanics.

Stochastic mechanics has been extended in various ways. In particular, the extension of stochastic mechanics to a single relativistic particle is now well understood [36, 53–60]. Also, the generalization of stochastic mechanics to particles moving on Riemannian manifolds [22, 61–65] is well established. Recently [66, 67], it was shown that the techniques developed in the stochastic mechanics literature to achieve such extensions can be defined in a rigorous way in the framework of second order geometry [68–71].

Furthermore, it has been pointed out that a description of spin lies within the scope of stochastic mechanics [22, 36, 61, 72–74], and stochastic mechanics has been extended to describe various multi-particle systems and field theories [22, 28, 36, 53, 53, 57, 58, 75–84]. In addition, various other aspects of stochastic mechanics have been studied, cf. e.g. Refs. [85–93]. Finally, as we will explain in Sect. 7.4, stochastic mechanics provides a mathematically clean description of the quantum foam that is hypothesized in quantum gravity. It is, therefore, not surprising that the tools from stochastic mechanics have also found their way into models of quantum gravity [94–98].

Despite the fact that stochastic mechanics provides a fully consistent description of quantum mechanics of a single particle, and has been used to explain many of the peculiar properties of quantum mechanics, the theory of stochastic mechanics has never become widely known. This may in part be ascribed to various unjustified criticisms of the theory [36]. Two of those will be discussed in more detail in Sects. 1.5 and 1.6.

A third criticism was formulated by Wallstrom [99, 100], which is discussed in more detail in Sect. 3.8. Various answers to this criticism have been formulated, cf. e.g. [101, 102], which require a new assumption in the theory of stochastic mechanics. In Sect. 3.8, we will show, however, that Wallstrom's criticism results from an incomplete analysis of stochastic mechanics. Therefore, the criticism is invalid and no new assumptions are necessary to resolve the criticism.

A fourth criticism was raised by Nelson [103, 104], and states that the processes that are studied in stochastic mechanics appear to fail to recover multi-time correlations predicted by quantum theories. A possible resolution of this criticism within the traditional framework of stochastic mechanics is discussed in Refs. [105, 106]. In this book, we adopt a different approach to this problem, as we modify the stochastic processes that are being studied. More precisely, we rotate the Wiener process in the complex plane and study its real projection. One can verify that these new processes correctly reproduce the multi-time correlations of quantum theories, and thus resolve Nelson's criticism. Using these new processes, the stochastic quantization condition is modified to the following statement:

- the trajectory of a stochastic particle can be described by a future and a past directed stochastic process $X(t) = C_\pm(t) + \text{Re}[M(t)]$;
- the processes $C_\pm(t)$ represent a drift that minimizes a stochastic action;
- the process $M(t)$ is a complex stochastic processes, such that $\text{Re}[M]$ and $\text{Im}[M]$ are correlated Wiener processes, i.e. it represents a Gaussian noise with covariance matrix $\mathbb{E}\left[M^i(t) M^j(t) \mid M(t_0)\right] = \alpha \frac{\hbar}{m} \delta^{ij} |t - t_0|$, where $\alpha \in \mathbb{C}$.

We conclude this section by emphasizing that stochastic mechanics is not in conflict with ordinary quantum theory. Instead, stochastic mechanics aims to derive the Dirac-Von Neumann axioms and the postulates of quantum mechanics from an underlying stochastic theory. Hence, any viable formulation of stochastic mechanics is consistent with all the results from standard quantum theory. The major difference is that stochastic mechanics is by construction compatible with the Kolmogorov axioms of probability theory. Therefore, in stochastic mechanics, the probability interpretation of quantum mechanics is naturally embedded in the theory, whereas in ordinary quantum theory the probability interpretation is imposed ad hoc by means of the Born rule [107]. Moreover, in the stochastic theory, the Hilbert space structure, that is central to the Dirac-Von Neumann axioms, arises in a natural way. Here, we will explain this in a qualitative fashion, making use of notions from measure theoretic probability theory that are reviewed in Appendices A and B.

Stochastic mechanics studies stochastic processes. By definition, a stochastic process is a family of random variables $\{X_t \mid t \in [t_0, t_f]\}$, where, for any $t \in [t_0, t_f]$, the random variable X_t is a map from a probability space $(\Omega, \Sigma, \mathbb{P})$ to the measurable configuration space $(\mathcal{M}, \mathcal{B}(\mathcal{M}))$. Hence, for any time $t \in [t_0, t_f]$, the stochastic process induces a measure $\mu_{X_t} = \mathbb{P} \circ X_t^{-1}$ on the configuration space $(\mathcal{M}, \mathcal{B}(\mathcal{M}))$. Then, in order to do analysis, one must introduce an L^p-norm that turns the space of all random variables into an L^p-space denoted by $L_t^p(\Omega, \Sigma, \mathbb{P})$.

The construction of the L^p-space of random variables allows to study the dynamics of the stochastic process, but does not provide the observables of the process. In a classical theory, observables are obtained by applying smooth functions $f \in C^\infty(T^*\mathcal{M})$ to the trajectory of the particle. In the stochastic theory, the observables are given by the expectation value of real valued measurable functions $f \in L^p(\mathcal{M}, \mathcal{B}(\mathcal{M}), \mu_{X_t})$ on the space $(\mathcal{M}, \mathcal{B}(\mathcal{M}))$ with induced measure μ_{X_t}. If one fixes the norm by setting $p = 2$, one finds that $L^2(\mathcal{M}, \mathcal{B}(\mathcal{M}), \mu_{X_t})$ is a Hilbert space. Therefore, in the stochastic theory, observables are elements of the real Hilbert space $L_\mathbb{R}^2(\mathcal{M}, \mathcal{B}(\mathcal{M}), \mu_{X_t})$.

This Hilbert space is different from the static Hilbert space $L_\mathbb{C}^2(\mathcal{M}, \mathcal{B}(\mathcal{M}), V_R)$, which is introduced in the Dirac-Von Neumann axioms, as it is real and time-dependent. However, by changing the measure from μ_{X_t} to the Riemann measure V_R, one obtains the real Hilbert space $L_\mathbb{R}^2(\mathcal{M}, \mathcal{B}(\mathcal{M}), V_R)$, and the Radon-Nykodým derivative associated to this change of measure is the probability density ρ_{X_t}. The complex Hilbert space $L_\mathbb{C}^2(\mathcal{M}, \mathcal{B}(\mathcal{M}), V_R)$ can then be obtained by complexifying $L_\mathbb{R}^2(\mathcal{M}, \mathcal{B}(\mathcal{M}), V_R)$. On this complex Hilbert space, the observables f act as self-adjoint operators and the elements of this Hilbert space are complex functions ψ, such that $|\psi|^2 \propto \rho_{X_t}$, where the proportionality denotes equality up to the scalar multiplication of the Hilbert space.

1.5 Time Reversibility

It is often believed that any stochastic process is inherently time irreversible. This idea has given rise to a criticism of any stochastic formulation of quantum mechanics that can be formulated as follows:

> Stochastic processes are often used to describe dissipative diffusion phenomena. These are non-unitary, thus time irreversible. Quantum mechanics, on the other hand, is a unitary theory, implying a degree of time reversibility. Therefore, quantum mechanics cannot be described by stochastic processes

At the heart of this criticism lies the incorrect assumption that any diffusion theory is non-unitary, hence inherently dissipative and time irreversible.[2] This assumption

---

[2] As pointed out in Ref. [22], the idea that dissipation is intrinsic is historically reminiscent to the Aristotelean school of thought. In Aristotelean dynamics, friction is a fundamental property of all physical systems, as it is believed that all objects tend to their rest position. For a long time, it was believed that this dynamical principle was correct, as it was compatible with the observations of the time, since most physical systems are subjected to friction. Galileo showed, however, that friction is not a fundamental property of deterministic dynamics. This led to Galilean relativity as a new dynamical principle in which friction is no longer fundamental to deterministic theories. This principle can be generalized to stochastic theories: although many stochastic systems are dissipative, dissipation is not a fundamental property of stochastic dynamics.

is itself based on a typical misunderstanding of the notion of time reversibility in stochastic theories, as we will explain below in a qualitative fashion.

The motion of a deterministic particle is governed by deterministic laws of motion. These are typically encoded in a Lagrangian or Hamiltonian formalism by means of a stationary action principle. Using this stationary action principle, one can derive equations of motion for the particle. These equations define an initial value problem, which allows to determine the trajectory $(X, V)(t)$ for all $t \in \mathcal{T} \subseteq \mathbb{R}$, when the state $(X, V)(t_0)$ is given at some $t_0 \in \mathcal{T}$.

A crucial aspect of deterministic theories is the principle of time reversal invariance, which states that the physical laws that govern the system must be invariant under time reversal. For deterministic theories, this time reversal invariance of the laws of motion implies that the solutions of the equations of motion have the same shape under a time reversal operation.

The notion of time reversibility is more subtle in stochastic theories. First of all, stochastic trajectories cannot be characterized completely by the state (X, V), due to the fact that the objects $\mathbb{E}[X^k] - \mathbb{E}[X]^k$, where $\mathbb{E}$ denotes expectation value and $k \in \mathbb{N}$, are non-vanishing in stochastic theories. Therefore, the state of a stochastic process is described by a state $(X, V_1, V_2, V_3, ...)$, where V_k denotes a velocity associated to the moment $\mathbb{E}[X^k]$. Only in deterministic theories, $\mathbb{E}[X^k] = \mathbb{E}[X]^k$, which implies that all velocities V_k are completely determined by V_1. This enables the phase space reduction to the state (X, V).

In order to simplify our further discussion, we will focus on processes for which all the moments $\mathbb{E}[X^k]$ are completely determined by the moments $\mathbb{E}[X]$ and $\mathbb{E}[X^2]$. For these processes, the state is described by a trajectory (X, V, V_2) in a higher order phase space, where V determines the drift velocity, while V_2 is the velocity of the variance $\text{Var}(X) = \mathbb{E}[X^2] - \mathbb{E}[X]^2$.

As is the case in a deterministic theory, the motion of the stochastic particle is governed by physical laws of motion. In particular, the deterministic laws, encoded in the stationary action principle, can be generalized to the stochastic theory by the construction of a stochastic Lagrangian or Hamiltonian. However, these laws will not completely determine the trajectory $(X, V, V_2)(t)$. Roughly speaking, these laws will only provide (X, V), while V_2 remains unknown. In order to obtain the full trajectory, one must introduce a new stochastic law of motion that fixes the velocity V_2.

To further simplify our discussion, we will now focus on the Wiener process. For the Wiener process localized at (x_0, t_0), the stochastic law of motion states that its evolution is governed by the heat kernel

$$\Phi(x, t; x_0, t_0) = \left[4\pi\alpha(t - t_0) \right]^{-\frac{d}{2}} \exp\left[-\frac{||x - x_0||^2}{4\alpha(t - t_0)} \right], \qquad (1.6)$$

where $\alpha > 0$ is the diffusion constant and d is the spatial dimension. This stochastic law only defines the forward evolution of the system, i.e. it defines the evolution for any $t, t_0 \in \mathcal{T}$ provided that $t \geq t_0$. In order to specify the backward evolution, we must impose another stochastic law, for which there exist infinitely many choices.

There is, however, a unique choice that is compatible with time reversal invariance, which is given by the kernel

$$\Phi(x, t; x_0, t_0) = \left[4\pi\alpha\,|t - t_0|\right]^{-\frac{d}{2}} \exp\left[-\frac{||x - x_0||^2}{4\alpha\,|t - t_0|}\right]. \qquad (1.7)$$

This new kernel defines the evolution of the process for any $t, t_0 \in \mathcal{T}$, and processes governed by this kernel are called two-sided Wiener processes [1].

By Lévy's characterization of the Wiener process [11], cf. Appendix B.8, imposing the heat kernel (1.6) is equivalent to imposing that the second order velocity is given by

$$V_2^{ij} = \alpha\,\delta^{ij}. \qquad (1.8)$$

Similarly, imposing the kernel (1.7) is equivalent to imposing

$$\begin{aligned} V_{2,+}^{ij} &= V_2^{ij} = \alpha\,\delta^{ij}, \\ V_{2,-}^{ij} &= -V_2^{ij} = -\alpha\,\delta^{ij}, \end{aligned} \qquad (1.9)$$

where the $+$ solution corresponds to processes evolving forward in time and the $-$ solution to processes evolving backward in time.

We emphasize that, by construction, the laws of motion, i.e. the stationary action principle and the stochastic law (1.9), are time reversal invariant. However, in contrast to the deterministic theory, this does not imply that the solutions of the equations of motion are also time reversal invariant. In fact, they are not, as the forward solution is described by $(X, V_+, +V_2)$, while a backward solution is given by $(X, V_-, -V_2)$, where[3] $V_+ \neq V_-$.

Often, in the study of stochastic processes, one only cares about the forward evolution of the system, such that one has to worry only about the stochastic law (1.8). Stochastic mechanics, on the other hand, studies processes for which the stochastic law is time reversible, which is imposed by setting $V_{2,-} = -V_{2,+}$. As a consequence, stochastic mechanics studies both the forward solutions (X, V_+, V_2) and the backward solutions $(X, V_-, -V_2)$.

Up to this point, we have discussed the notion of time reversibility in stochastic theories. This discussion does not resolve the criticism formulated at the beginning of this section, as the two-sided Wiener process governed by the kernel (1.7) remains a dissipative process, albeit with respect to both the future and the past. However, things change drastically, when one considers superpositions of processes. Stochastic mechanics proves that such superpositions can be governed by a unitary time evolution, as encountered in quantum mechanics.

---

[3] The presence of two different velocities V_+ and V_- reflects that a Wiener process is almost surely not differentiable. This can be interpreted as follows: at every time $t \in \mathcal{T}$, the Brownian particle gets a kick from a microscopic particle, which induces a discontinuous change of velocity from V_- to V_+.

The traditional approach in the stochastic mechanics literature, cf. e.g. Refs. [1, 21, 22, 36, 52], for obtaining such processes with a unitary time evolution is to consider a superposition of the forward and backward solutions. As mentioned in Sect. 1.4, we adopt a different approach in this book: we consider two correlated two-sided Wiener processes M_x and M_y, and study the complex process $M = M_x + i M_y$. Then, we show that the evolution of the real projection $\text{Re}[M] = M_x$ is governed by the heat kernel (1.7), if the processes M_x and M_y are uncorrelated, and that the evolution is governed by a unitary Schrödinger kernel, if the processes are maximally correlated.

1.6 Hidden Variables

Another criticism that is often faced by stochastic interpretations of quantum mechanics is that stochastic theories open the door to hidden variable theories. This criticism can be formulated as follows

> Stochastic theories often arise as an effective theory that replaces a more fundamental hidden background field of microscopic particles. On the other hand, the Bell theorems exclude any locally real hidden variable theory. Therefore, quantum theories must be fundamentally different from stochastic theories.

In the physical models of Brownian motion and other stochastic processes, the theory contains a background field and expectation values are obtained by calculating ensemble averages. This background field is governed by deterministic laws of motion, but, due to the enormous computational complexity of this deterministic theory, one introduces an effective stochastic theory. This stochastic theory allows to determine the behavior of the macroscopic particle in a much more efficient manner.

In the mathematical theory of stochastic processes, the background field is replaced by a probability space and expectation values are defined as Lebesgue integrals over this probability space. Stochastic mechanics is built entirely within this mathematical framework. Therefore, stochastic mechanics should be interpreted as a mathematically rigorous implementation of the statement that 'God plays dice'.

Stochastic mechanics remains agnostic about the question whether the stochastic theory based on the idea of 'God playing dice' may be replaced with a statistical theory containing some background field. Moreover, if such a background field is introduced, stochastic mechanics only imposes a condition on the stochastic law that is obtained in the continuum limit.[4] It does not impose any conditions on the (non-)deterministic laws that govern the background field on a microscopic level.

We point out that the Bell theorems do not rule out the replacement of the probability space in stochastic mechanics by some background field that is governed by

---

[4] The condition is that the ensemble averages $\langle . \rangle$ calculated in the statistical theory converge to the expectation values $\mathbb{E}[.]$ calculated in the stochastic theory, i.e., for any observable A, $\langle A \rangle \to \mathbb{E}[A]$ in the limit where the number of particles in the background field goes to infinity, while their size goes to zero.

deterministic laws of motion. Such deterministic hidden variable theories generically imply the presence of uncertainty principles in the corresponding stochastic theory, while the derivation of the Bell inequalities relies on the absence of such uncertainty relations. This argument is worked out in more detail in Sect. 7.3.

1.7   Outline for the Book

In this book, we study a spinless stochastic particle subjected to a scalar potential $\mathfrak{U}$ and a vector potential A. We derive the stochastic equations of motion for this particle and show that the (proper) time evolution of the probability density of this particle is subjected to the diffusion equation (1.1) with $\mathcal{M}$ a Riemannian or Lorentzian manifold, $\Psi \in L^2(\mathcal{M})$ complex valued, and H given by the covariant generalization of Eq. (1.2).

Conversely, we show, by explicit construction, that for any solution of the diffusion equation (1.1) with $\mathcal{M}$ a Riemannian or Lorentzian manifold, $\Psi \in L^2(\mathcal{M})$, and H given by the covariant generalization of Eq. (1.2), there exists a stochastic process X such that $|\Psi|^2$ describes the probability density of X.

The presentation in this book is self-contained: the results review, reformulate and build on results from the stochastic mechanics literature, but no prior knowledge of stochastic mechanics is assumed.

The book is organized as follows: in Chap. 2, we review the classical dynamics of a deterministic spinless non-relativistic particle on $\mathbb{R}^d$. In Chap. 3, we superimpose a Brownian motion onto this particle and derive the Feynman-Kac theorem using the tools from stochastic mechanics. In Chap. 4, we complexify the Wiener process and perform the same analysis, which allows to extend the results from real to complex diffusion equations. In Chap. 5, we extend this complex stochastic theory to relativistic particles on $\mathbb{R}^{d,1}$. In Chap. 6, we extend the results to Riemannian and Lorentzian manifolds. In Chap. 7, we discuss some important aspects of the stochastic interpretation of quantum mechanics that is implied by the results. Finally, in Chap. 8, we conclude.

Our analysis heavily relies on standard results from stochastic analysis, which are reviewed in Appendices A, B and C. Moreover, the extension of the theory to manifolds requires the framework of second order geometry, which is reviewed in Appendix D. Finally, in Appendices E and F, the equations of motion for the stochastic particle are derived using stochastic variational calculus.

Chapter 2
Classical Dynamics on $\mathbb{R}^d$

Abstract This chapter concisely reviews the Lagrangian and Hamiltonian formulations of classical mechanics of a single particle.

We consider a particle with mass m moving in the d-dimensional real space $\mathbb{R}^d$. We assume the particle to be charged under a vector potential $A_i(x, t)$ with charge q and a scalar potential $\mathfrak{U}(x, t)$. We are interested in its trajectory $\{X_t \mid t \in \mathcal{T}\}$ with $X_t = X(t) : \mathcal{T} \to \mathbb{R}^d$ parameterized by the time $t \in \mathcal{T} = [t_0, t_f]$ with $t_0 < t_f \in \mathbb{R}$.

The motion of this particle is governed by a Lagrangian $L : T\mathbb{R}^d \times \mathcal{T} \to \mathbb{R}$ that is defined on the tangent bundle (phase space) $T\mathbb{R}^d \cong \mathbb{R}^{2d}$ and is given by

$$L(x, v, t) = \frac{m}{2} \, \delta_{ij} \, v^i v^j + q \, A_i(x, t) \, v^i - \mathfrak{U}(x, t) \,. \tag{2.1}$$

The corresponding action is given by the integral

$$S(X) = \int_{t_0}^{t_f} L(X_t, V_t, t) \, dt \,, \tag{2.2}$$

where $(X, V) : \mathcal{T} \to T\mathbb{R}^d$ describes a trajectory on the tangent bundle. By extremizing this action one finds the Euler-Lagrange equations. These are given by

$$\frac{d}{dt} \frac{\partial L}{\partial v^i} = \left. \frac{\partial L}{\partial x^i} \right|_{(x,v)=(X_t, V_t)} \,. \tag{2.3}$$

For the Lagrangian (2.1), the Euler-Lagrange equations yield

$$\frac{d}{dt} \left(m \, \delta_{ij} \, V^j + q \, A_i \right) = q \, V^j \partial_i A_j - \partial_i \mathfrak{U} \,. \tag{2.4}$$

Then, using that the velocity satisfies

$$V_t^i = \frac{dX_t^i}{dt} \,, \tag{2.5}$$

F. Kuipers, *Stochastic Mechanics*, SpringerBriefs in Physics,
https://doi.org/10.1007/978-3-031-31448-3_2

one finds

$$m\,\delta_{ij}\frac{dV^j}{dt} = q\,F_{ij}\,V^j - q\,\partial_t A_i - \partial_i \mathfrak{U},\tag{2.6}$$

where the field strength is defined by

$$F_{ij} := \partial_i A_j - \partial_j A_i.\tag{2.7}$$

The trajectory $(X, V) : \mathcal{T} \to T\mathbb{R}^d$ is now uniquely determined by Eqs. (2.5) and (2.6) supplemented with an initial condition, i.e. by the initial value problem

$$\begin{cases} \frac{dX_t^i}{dt} &= V_t^i,\\[4pt] m\,\delta_{ij}\frac{dV_t^j}{dt} &= q\,F_{ij}(X_t, t)\,V_t^j - q\,\partial_t A_i(X_t, t) - \partial_i \mathfrak{U}(X_t, t),\\[4pt] (X_{t_0}, V_{t_0}) &= (x_0, v_0). \end{cases}\tag{2.8}$$

Alternatively, the trajectory of the particle can be derived in the Hamiltonian formulation of classical mechanics. In this equivalent description, the motion is governed by a Hamiltonian $H : T^*\mathbb{R}^d \times \mathcal{T} \to \mathbb{R}$ defined on the cotangent bundle (phase space) $T^*\mathbb{R}^d \cong \mathbb{R}^{2d}$, which can be obtained from the Lagrangian by a Legendre transform

$$H(x, p, t) = p_i v^i - L(x, v, t),\tag{2.9}$$

where p_i is the canonical momentum defined by

$$p_i := \frac{\partial L}{\partial v^i}.\tag{2.10}$$

Inversely, the Lagrangian can be obtained from the Hamiltonian through a Legendre transform

$$L(x, v, t) = p_i v^i - H(x, p, t)\tag{2.11}$$

with the canonical velocity

$$v^i := \frac{\partial H}{\partial p_i}.\tag{2.12}$$

Using the Hamiltonian, one can derive the Hamilton equations, which is a set of ordinary differential equations. These are equivalent to the Euler-Lagrange equations and given by

$$\begin{cases} \frac{dx^i}{dt} &= \frac{\partial H}{\partial p_i}\\[4pt] \frac{dp_i}{dt} &= -\frac{\partial H}{\partial x^i}\\[4pt] \frac{\partial H}{\partial t} &= -\frac{\partial L}{\partial t} \end{cases}\Bigg|_{(x,p)=(X_t, P_t)}.\tag{2.13}$$

There exists a third equivalent formulation of classical mechanics, which is the Hamilton-Jacobi formalism. In this formulation, one defines Hamilton's principal function $S : \mathbb{R}^d \times \mathcal{T} \to \mathbb{R}$, such that

$$S(x, t) = S(x, t; x_0, t_0) = \int_{t_0}^{t} L(X_s, V_s, s)\, ds \,, \tag{2.14}$$

where (X_s, V_s) is a solution of the Euler-Lagrange equations passing through (x_0, t_0) and (x, t) with $t_0 \leq t \leq t_f$. Using Hamilton's principal function, one can derive another equivalent set of equations of motion. These are the Hamilton-Jacobi equations:

$$\begin{cases} \frac{\partial S(x,t)}{\partial x^i} = p_i \\ \frac{\partial S(x,t)}{\partial t} = -H(x, p, t) \end{cases} \Bigg|_{(x,p)=(X_t, p(X_t,t))}. \tag{2.15}$$

For the Lagrangian (2.1), this yields

$$\begin{cases} \frac{\partial S}{\partial x^i} = m\, \delta_{ij} v^j + q\, A_i \\ \frac{\partial S}{\partial t} = -\frac{m}{2}\, \delta_{ij} v^i v^j - \mathfrak{U} \end{cases} \Bigg|_{(x,v)=(X_t, v(X_t,t))}. \tag{2.16}$$

Here, (x, v) and (x, p) are coordinates on the (co)tangent bundle, $v(x, t)$ and $p(x, t)$ are (co)vector fields over $\mathbb{R}^d$, X_t is a trajectory on $\mathbb{R}^d$ and (X_t, V_t), (X_t, P_t) are trajectories on the (co)tangent bundle.

One can explicitly show the equivalence between the Hamilton-Jacobi equations (2.16) and the Euler-Lagrange equations (2.6) by taking a spatial derivative of the second Hamilton-Jacobi equation and plugging in the first equation. This leads to

$$m\, \delta_{ij} \frac{\partial}{\partial t} v^j + q\, \frac{\partial}{\partial t} A_i = -m\, \delta_{jk} v^k \partial_i v^j - \partial_i \mathfrak{U}. \tag{2.17}$$

Then, using the first Hamilton-Jacobi equation, one finds

$$\begin{aligned} m\, \delta_{jk}\, \partial_i v^j &= \partial_i\, (\partial_k S - q\, A_k) \\ &= \partial_k\, (\partial_i S - q\, A_i) - q\, (\partial_i A_k - \partial_k A_i) \\ &= m\, \delta_{ij}\, \partial_k v^j - q\, F_{ik}\,. \end{aligned} \tag{2.18}$$

Plugging this relation into Eq. (2.17) yields

$$\left[m\, \delta_{ij} \left(\frac{\partial}{\partial t} + v^k \partial_k \right) - q\, F_{ij} \right] v^j = -q\, \frac{\partial}{\partial t} A_i - \partial_i \mathfrak{U}. \tag{2.19}$$

Finally, using that

$$\frac{dX_t}{dt} = v(X_t, t),$$ (2.20)

one finds that this is equivalent to Eq. (2.6) with V_t replaced by $v(X_t, t)$.

We conclude this chapter with two remarks. First, we note that one can write down a partial differential equation for Hamilton's principal function by combining the expressions in Eq. (2.16). This yields

$$- 2m \frac{\partial S}{\partial t} = \partial^i S \, \partial_i S - 2q \, A^i \, \partial_i S + q^2 \, A^i A_i + 2m \, \mathfrak{U}.$$ (2.21)

Secondly, we point out that Hamilton's principal function can also be defined as

$$S(x, t) = S(x, t; x_f, t_f) = - \int_t^{t_f} L(X_s, V_s, s) \, ds,$$ (2.22)

where (X_s, V_s) is a solution of the Euler-Lagrange equations passing through (x, t) and (x_f, t_f) with $t_0 \le t \le t_f$. The Hamilton-Jacobi equations for this principal function are also given by Eq. (2.15).

Chapter 3
Stochastic Dynamics on $\mathbb{R}^d$

Abstract This chapter discusses the theory of Brownian motion (Wiener process) with drift in detail. It derives the stochastic differential equations of motion from a stochastic action principle in the Lagrangian formulation. It shows how the Feynman-Kac formula can be derived from the stochastic Hamilton-Jacobi equation. Moreover, it discusses some important differences between deterministic theories and stochastic theories.

In previous chapter, we discussed the equations of motion that govern a deterministic theory. In this chapter, we introduce a notion of randomness in the theory by promoting the deterministic trajectories to stochastic processes. We will be interested in processes X of the form

$$X_t = C_t + M_t , \qquad (3.1)$$

where C represents a deterministic trajectory, while M represents a stochastic noise. Our aim is to derive equations of motion for such a process by minimizing an action, as was done in the previous chapter.

We can make this idea precise in the language of stochastic analysis, which we review in Appendices A and B. In this language, a stochastic process that can be decomposed as in Eq. (3.1) is called a semi-martingale, the process C is a càdlàg process[1] of finite variation[2] and M is a local martingale process.[3]

In addition, we will impose a notion of time reversibility to the process by requiring that the laws of motion are invariant under time reversal. As discussed in Sect. 1.5, the time reversal symmetry of the laws of motion does not imply time reversal symmetry

---

[1] A càdlàg process is right-continuous with left limits. In this book, starting from Eq. (3.17), we only consider processes that are continuous, which is a stronger assumption than being càdlàg.

[2] Roughly speaking, the requirement of finite variation ensures that C is deterministic, bounded and that it does not oscillate with an infinite frequency.

[3] A martingale process is a process that does not drift, i.e. its expectation value is constant in time: $\mathbb{E}[M_t | \{M_r : r \in [t_0, s]\}] = M_s$ for all $t > s \in \mathcal{T}$. For local martingales, this martingale property is only required to hold locally, i.e. if s and t are close to each other.

© The Author(s), under exclusive license to Springer Nature Switzerland AG 2023
F. Kuipers, *Stochastic Mechanics*, SpringerBriefs in Physics,
https://doi.org/10.1007/978-3-031-31448-3_3

of the solutions of the equations of motion. Therefore, solutions will be a two-sided[4] semi-martingale of the form

$$X_t^i = C_{\pm,t}^i + M_t^i , \tag{3.2}$$

where C_+ is càdlàg process of finite variation that evolves forward in time and C_- is càglàd process[5] of finite variation that evolves backward in time. Moreover, M is a two-sided local martingale.

In the following chapters, it will prove to be useful to let the processes X and M evolve in different copies of the configuration space $\mathbb{R}^d$. This can be achieved by rewriting Eq. (3.2) as

$$X_t^i = C_{\pm,t}^i + \delta_a^i M_t^a . \tag{3.3}$$

Here, we have promoted the configuration space $\mathcal{M} = \mathbb{R}^d$ to the frame bundle $(E, \pi, \mathcal{M})$ with base space $\mathcal{M} = \mathbb{R}^d$ and fibers $F = \mathbb{R}^d$, such that X is a two-sided process on the base space $\mathcal{M}$, $C_\pm$ are one-sided processes on the base space $\mathcal{M}$, and M is a two-sided martingale on the fibers F. Moreover, the Kronecker delta δ_a^i defines an orthonormal frame spanning F at every point $x \in \mathcal{M}$ and is commonly referred to as a vielbein or polyad.

In the remainder of this chapter, we will discuss the derivation of the equations of motion for the stochastic trajectory by generalizing the principle of stationary action to a stochastic theory. Such equations of motion will only fix the deterministic part contained in C, but not the noise contained in M. Therefore, we must impose additional laws of motion to fix the stochastic behavior of X.

3.1 The Stochastic Law

The aim of this section is to fix the stochastic law of the martingale process M. In this chapter, we will fix M to be a Wiener process. However, as we will generalize the theory in the next chapters, we will provide a more general discussion.

A standard way of fixing the stochastic law of a stochastic process X is by fixing its characteristic functional

$$\varphi_X(J) := \mathbb{E}\left[e^{i \int_T J_i(t) X^i(t) \, dt} \right]. \tag{3.4}$$

Our first assumption is that the moment generating functional for X, defined by

$$M_X(J) := \mathbb{E}\left[e^{\int_T J_i(t) X^i(t) \, dt} \right], \tag{3.5}$$

---

[4] Being two-sided means that the stochastic evolution of the process is well defined and governed by the same stochastic law with respect to both the future and past directed evolution.

[5] A càglàd process is left-continuous with right limits.

has a non-zero radius of convergence. Due to this assumption, fixing $\varphi_X(J)$ is equivalent to fixing $M_X(J) = \varphi_X(-i\,J)$. In addition, the non-zero radius of convergence of $M_X(J)$ implies that all moments $\mathbb{E}[X^k]$ for $k \in \mathbb{N}$ exist, and that, within its radius of convergence, $M_X(J)$ can be characterized completely by its moments. Therefore, we can determine the stochastic law of X by specifying the moments $\mathbb{E}[X^k]$ for all $k \in \mathbb{N}$. Furthermore, since $X = C_\pm + M$ and $C_\pm$ has finite variation, $\mathbb{E}[C_\pm^k] = \mathbb{E}[C_\pm]^k$. Hence, we only need to specify the moments $\mathbb{E}[M^k]$, More precisely, we must specify all moments

$$\mathbb{E}\left[\prod_{i=1}^{k}(M_{t_i}^{a_i} - M_{s_i}^{a_i})\right] \qquad \forall\, t_i, s_i \in \mathcal{T},\ a_i \in \{1, ..., d\},\ k \in \mathbb{N}. \qquad (3.6)$$

Since M is a martingale, the first moment is fixed by the martingale property (B.4), as

$$\begin{aligned}
\mathbb{E}[(M_t^a - M_s^a)] &= \mathbb{E}[\mathbb{E}[(M_t^a - M_s^a)|\{M_r : t_0 \leq r \leq s\}]] \\
&= \mathbb{E}[M_s^a - M_s^a] \\
&= 0\,.
\end{aligned} \qquad (3.7)$$

In order to fix the other moments, we will make the additional assumption that M is a Lévy process.[6] Therefore, M has independent increments, and, due to this independence,[7] all moments (3.6) are completely determined by the subset of moments

$$\mathbb{E}\left[\prod_{i=1}^{k}(M_t^{a_i} - M_s^{a_i})\right] \qquad \forall\, s < t \in \mathcal{T},\ a_i \in \{1, ..., d\},\ k \in \mathbb{N}. \qquad (3.8)$$

We will fix these remaining moments by imposing a structure relation, using a differential notation. In this differential notation, the decomposition (3.3) is given by the stochastic differential equation[8]

$$d_\pm X_t^i = v_\pm^i(X_t, t)\, dt + \delta_a^i\, d_\pm M_t^a\,, \qquad (3.9)$$

where d_+ is a forward Itô differential and d_- is a backward Itô differential, i.e.

$$\begin{aligned}
d_+ X_t &:= X_{t+dt} - X_t\,, \\
d_- X_t &:= X_t - X_{t-dt}\,.
\end{aligned} \qquad (3.10)$$

---

[6] Lévy processes can be regarded as the continuous time analogue of the random walk, cf. Appendix B.7. Examples of Lévy processes include the Wiener process and the Poisson process.

[7] Another consequence of the independent increments is that M is a Markov process.

[8] Cf. Appendix C for a review of stochastic calculus.

We introduce a bracket $[.,.]$, called the quadratic variation[9] of X, which is given by

$$d_+[X, X]_t := \left[X_{t+dt} - X_t\right] \otimes \left[X_{t+dt} - X_t\right],$$
$$d_-[X, X]_t := \left[X_t - X_{t-dt}\right] \otimes \left[X_t - X_{t-dt}\right]. \tag{3.11}$$

Since $C_\pm$ has finite variation, $dC_t = \mathcal{O}(dt)$, but $dM_t = o(1)$, whence[10]

$$d_\pm[X, X]_t = d_\pm[C, C]_t + d_\pm[C, M]_t + d_\pm[M, C]_t + d_\pm[M, M]_t$$
$$= d_\pm[M, M]_t + o(dt). \tag{3.12}$$

The quadratic variation can be used to calculate the quadratic moment of M, since

$$\mathbb{E}\left[(M_t^a - M_s^a)(M_t^b - M_s^b)\right] = \mathbb{E}\left[\mathbb{E}\left[(M_t^a - M_s^a)(M_t^b - M_s^b)\,\middle|\,\{M_r : t_0 \leq r \leq s\}\right]\right]$$
$$= \mathbb{E}\left[\int_s^t d_+M_{r_1}^a \int_s^t d_+M_{r_2}^b\right]$$
$$= \mathbb{E}\left[\int_s^t d_+[M^a, M^b]_r\right], \tag{3.13}$$

where we used that $M_t = M_s + \int_s^t d_+M_r$ for $s < t$. Furthermore, using that $M_s = M_t + \int_t^s d_-M_r$ for $s < t$, we find

$$\mathbb{E}\left[(M_t^a - M_s^a)(M_t^b - M_s^b)\right] = \mathbb{E}\left[\mathbb{E}\left[(M_t^a - M_s^a)(M_t^b - M_s^b)\,\middle|\,\{M_r : t \leq r \leq t_f\}\right]\right]$$
$$= \mathbb{E}\left[\int_t^s d_-M_{r_1}^a \int_t^s d_-M_{r_2}^b\right]$$
$$= \mathbb{E}\left[\int_t^s d_-[M^a, M^b]_r\right]$$
$$= \mathbb{E}\left[-\int_s^t d_-[M^a, M^b]_r\right]. \tag{3.14}$$

Now, we can introduce the notation

$$d[M, M]_t := d_+[M, M]_t = -d_-[M, M]_t, \tag{3.15}$$

and impose a structure relation for the quadratic variation $d[M, M]_t$. A general structure relation takes the form

$$d[M^a, M^b] = \frac{\alpha\,\hbar}{m} A^{ab}\,dt + \frac{\beta}{\kappa} B_c^{ab}\,dM_t^c, \tag{3.16}$$

---

[9] Cf. Appendix B.6 for a definition and discussion of the properties of quadratic variation.

[10] This reflects the fact that the stochastic law of X is determined by the stochastic law of M.

where A and B are dimensionless structure constants with A symmetric and positive definite. m, $\hbar$ and κ are introduced to fix the physical dimensions, such that m is the mass of the particle, $\hbar$ the reduced Planck constant and κ a constant with dimension $[\kappa] = L^{-1}$. $\alpha \in [0, \infty)$ and $\beta \in \mathbb{R}$ are dimensionless parameters and α, $\beta \to 0$ yield the deterministic limit. Furthermore, M is continuous, if and only if $\beta = 0$.

In this book, we assume that X is continuous, which fixes $\beta = 0$, we set $A^{ab} = \delta^{ab}$ and work in natural units where $\hbar = 1$. The structure relation then simplifies to

$$m \, d[M^a, M^b]_t = \alpha \, \delta^{ab} \, dt \qquad (3.17)$$

with[11] $\alpha \geq 0$. We note that, by the Lévy characterization of Brownian motion, this structure relation implies that M is a Wiener process, cf. Appendix B.8.

The structure relation (3.17) fixes all the moments given in Eq. (3.8). Indeed, using Eq. (3.13), we find that the quadratic moment is given by

$$
\begin{aligned}
\mathbb{E}\left[(M_t^a - M_s^a)(M_t^b - M_s^b)\right] &= \mathbb{E}\left[\int_s^t d[M^a, M^b]_r\right] \\
&= \mathbb{E}\left[\frac{\alpha}{m} \int_s^t \delta^{ab} \, dr\right] \\
&= \frac{\alpha}{m}(t - s)\, \delta^{ab} .
\end{aligned}
\qquad (3.18)
$$

Furthermore, all moments of order k can be expressed as a linear combination of moments of order $(k - 2)$ using that

$$
\begin{aligned}
\mathbb{E}\left[\prod_{i=1}^k (M_t^{a_i} - M_s^{a_i})\right] &= \mathbb{E}\left[\prod_{i=1}^k \int_s^t dM_{r_i}^{a_i}\right] \\
&= \sum_{j=1}^{k-1} \mathbb{E}\left[\int_s^t d[M^{a_k}, M^{a_j}]_{r_k} \prod_{i=1, i\neq j}^{k-1} \int_s^t dM_{r_i}^{a_i}\right] \\
&= \frac{\alpha}{m}(t - s) \sum_{j=1}^{k-1} \delta^{a_k a_j} \mathbb{E}\left[\prod_{i=1, i\neq j}^{k-1} (M_t^{a_i} - M_s^{a_i})\right].
\end{aligned}
\qquad (3.19)
$$

Therefore,

$$\mathbb{E}\left[\prod_{i=1}^k (M_t^{a_i} - M_s^{a_i})\right] = 0 \qquad \text{if } k \text{ is odd} \qquad (3.20)$$

and all even moments can be expressed in terms of the quadratic moment (3.18). We remark that this result is known as Isserlis' theorem or Wick's probability theorem.

---

[11] Note that the real martingale M exists, if and only if $\alpha \geq 0$, and is non-trivial, if and only if $\alpha \neq 0$.

3.2 Stochastic Phase Space

Now that we have fixed the stochastic law of X, we would like to derive the equations of motion for X. In order to do so, we must construct a stochastic action using a stochastic Lagrangian and derive equations of motion by stochastically minimizing this action.

Here, we encounter a difficulty: the classical Lagrangian is defined on the tangent bundle $T\mathbb{R}^d \cong \mathbb{R}^{2d}$, so we expect that the stochastic Lagrangian is also defined on a tangent bundle. However, stochastic processes are almost surely not differentiable. Therefore, there is no trivial notion of velocity. In the deterministic theory, we can define for any trajectory X a velocity, as given in Eq. (2.5):

$$
\begin{aligned}
V_t &:= \lim_{dt \to 0} \frac{X_{t+dt} - X_{t-dt}}{2\,dt} \\
&= \lim_{dt \to 0} \frac{X_{t+dt} - X_t}{dt} \\
&= \lim_{dt \to 0} \frac{X_t - X_{t-dt}}{dt},
\end{aligned}
\tag{3.21}
$$

such that the pair (X_t, V_t) is a trajectory on the tangent bundle $T\mathbb{R}^d \cong \mathbb{R}^{2d}$. However, when X is a stochastic process, none of these expressions is well defined.

Nevertheless, using conditional expectations, we can construct a velocity field, which is the stochastic equivalent of the velocity field given in Eq. (2.20), but there exist two inequivalent possibilities:

$$
v_+(X_t, t) = \lim_{dt \to 0} \mathbb{E}\left[\frac{X_{t+dt} - X_t}{dt} \,\middle|\, X_t \right],
\tag{3.22}
$$

$$
v_-(X_t, t) = \lim_{dt \to 0} \mathbb{E}\left[\frac{X_t - X_{t-dt}}{dt} \,\middle|\, X_t \right],
\tag{3.23}
$$

which are the forward and backward Itô velocities. In addition, we can define a Stratonovich velocity by

$$
\begin{aligned}
v_\circ(X_t, t) &= \lim_{dt \to 0} \mathbb{E}\left[\frac{X_{t+dt} - X_{t-dt}}{2\,dt} \,\middle|\, X_t \right] \\
&= \lim_{dt \to 0} \left\{ \mathbb{E}\left[\frac{X_{t+dt} - X_t}{2\,dt} \,\middle|\, X_t \right] + \mathbb{E}\left[\frac{X_t - X_{t-dt}}{2\,dt} \,\middle|\, X_t \right] \right\} \\
&= \frac{1}{2}\left[v_+(X_t, t) + v_-(X_t, t) \right].
\end{aligned}
\tag{3.24}
$$

Moreover, in contrast to the classical case, we can define another non-vanishing velocity field, which results from the non-vanishing quadratic variation and is given by

$$v_2(X_t, t) := \lim_{dt \to 0} \mathbb{E}\left[\frac{(X_{t+dt} - X_t) \otimes (X_{t+dt} - X_t)}{dt} \,\bigg|\, X_t\right]. \qquad (3.25)$$

One could again take the limit in various ways, yielding velocities $v_{2,+}$, $v_{2,-}$ and $v_{2,\circ}$. However, this does not lead to independent velocities, since, by Eqs. (3.13) and (3.14),

$$v_{2,+}(X_t, t) = v_2(X_t, t),$$
$$v_{2,-}(X_t, t) = -v_2(X_t, t),$$
$$v_{2,\circ}(X_t, t) = 0. \qquad (3.26)$$

In the classical theory, the velocity fields are sections of the tangent bundle $T\mathbb{R}^d \cong \mathbb{R}^{2d}$. Similarly, in the stochastic theory, the fields $v_\circ$, (v_+, v_2), $(v_-, -v_2)$ can be regarded as sections of the tangent bundles $T_\circ\mathbb{R}^d$, $T_+\mathbb{R}^d$ and $T_-\mathbb{R}^d$ respectively.[12] For the Stratonovich bundle we have $T_\circ\mathbb{R}^d \cong \mathbb{R}^{2d}$, but the Itô bundles $T_\pm\mathbb{R}^d$ have a larger dimension, due to the $\frac{d(d+1)}{2}$ additional degrees of freedom contained in the symmetric object v_2, such that $T_\pm\mathbb{R}^d \cong \mathbb{R}^{\frac{d(d+5)}{2}}$.

We can study processes $(X_t, V_{\circ,t})$ on $T_\circ\mathbb{R}^d$ and processes $(X_t, V_{\pm,t}, \pm V_{2,t})$ on $T_\pm\mathbb{R}^d$ and define a relation between X_t and V_t, which is similar to the classical relation given in Eq. (2.5): the processes $V_\circ$, V_+, V_- and V_2 are the velocity above X, if

$$\mathbb{E}\left[\int f_i(X_t, t)\, V^i_{\circ,t}\, dt\right] = \mathbb{E}\left[\int f_i(X_t, t)\, d_\circ X^i_t\right],$$

$$\mathbb{E}\left[\int f_i(X_t, t)\, V^i_{+,t}\, dt\right] = \mathbb{E}\left[\int f_i(X_t, t)\, d_+ X^i_t\right],$$

$$\mathbb{E}\left[\int f_i(X_t, t)\, V^i_{-,t}\, dt\right] = \mathbb{E}\left[\int f_i(X_t, t)\, d_- X^i_t\right],$$

$$\mathbb{E}\left[\int g_{ij}(X_t, t)\, V^{ij}_{2,t}\, dt\right] = \mathbb{E}\left[\int g_{ij}(X_t, t)\, d[X^i, X^j]_t\right] \qquad (3.27)$$

for any Lebesgue integrable $f \in T^*\mathbb{R}^d$ and $g \in T^2(T^*\mathbb{R}^d)$. In the remainder of the book, we will shorten these expressions using the differential notation

$$V^i_{\circ,t}\, dt = d_\circ X^i_t, \qquad (3.28)$$
$$V^i_{+,t}\, dt = d_+ X^i_t, \qquad (3.29)$$
$$V^i_{-,t}\, dt = d_- X^i_t, \qquad (3.30)$$
$$V^{ij}_{2,t}\, dt = d[X^i, X^j]_t. \qquad (3.31)$$

---

[12] Cf. Appendix D for more detail.

The velocity field v_2 is fixed by the structure relation for M, as the structure relation (3.17) implies

$$
\begin{aligned}
m\, d[X^i, X^j]_t &= \delta_a^i \delta_b^j \, d[M^a, M^b]_t + o(dt) \\
&= \alpha\, \delta_a^i \delta_b^j \delta^{ab}\, dt + o(dt) \\
&= \alpha\, \delta^{ij}\, dt + o(dt) .
\end{aligned}
\tag{3.32}
$$

Hence,

$$
\begin{aligned}
v_2^{ij} &= \lim_{dt \to 0} \mathbb{E}\left[\frac{d[X^i, X^j]}{dt} \right] \\
&= \frac{\alpha}{m}\, \delta^{ij} .
\end{aligned}
\tag{3.33}
$$

3.3 Stochastic Action

In the previous section, we have constructed the stochastic phase spaces. This allows to define Lagrangian functions $L^\circ : T_\circ \mathbb{R}^d \times \mathcal{T} \to \mathbb{R}$, $L^\pm : T_\pm \mathbb{R}^d \times \mathcal{T} \to \mathbb{R}$ and action functionals

$$
S_\circ(X) = \mathbb{E}\left[\int_{t_0}^{t_f} L^\circ(x, v_\circ, t)\, dt \right],
\tag{3.34}
$$

$$
S_\pm(X) = \mathbb{E}\left[\int_{t_0}^{t_f} L^\pm(x, v_\pm, v_2, t)\, dt \right].
\tag{3.35}
$$

We must now find the stochastic equivalent of the classical Lagrangian (2.1). Since the Stratonovich tangent bundle is similar to the classical tangent bundle, there is a natural choice for the Stratonovich Lagrangian:

$$
L^\circ(x, v_\circ, t) = L(x, v_\circ, t) .
\tag{3.36}
$$

Hence, the Stratonovich Lagrangian associated with the classical Lagrangian (2.1) is

$$
L^\circ(x, v_\circ, t) = \frac{m}{2}\, \delta_{ij}\, v_\circ^i v_\circ^j + q\, A_i(x, t)\, v_\circ^i - \mathfrak{U}(x, t) .
\tag{3.37}
$$

For the Itô Lagrangians, on the other hand, there is no obvious choice. We can, however, construct these Lagrangians from the Stratonovich Lagrangian by imposing

$$
S(X) := S_\circ(X) = S_\pm(X) .
\tag{3.38}
$$

In Appendix E, we show that this condition implies that the forward and backward Itô Lagrangians corresponding to the Stratonovich Lagrangian (3.37) are given by

$$L^{\pm}(x, v_{\pm}, v_2, t) = L_0^{\pm}(x, v_{\pm}, v_2, t) \pm L_\infty(x, v_\circ) \tag{3.39}$$

with finite part

$$L_0^{\pm}(x, v_{\pm}, v_2, t) = \frac{m}{2} \delta_{ij} v_{\pm}^i v_{\pm}^j + q\, A_i(x, t)\, v_{\pm}^i \pm \frac{q}{2} \partial_j A_i(x, t)\, v_2^{ij} - \mathfrak{U}(x, t) \tag{3.40}$$

and a divergent part defined by the integral condition

$$\mathbb{E}\left[\int L_\infty(x, v_\circ)\, dt\right] = \mathbb{E}\left[\int \frac{m}{2} \delta_{ij}\, d[x^i, v_\circ^j]\right]. \tag{3.41}$$

3.4 Stochastic Euler-Lagrange Equations

By minimizing the stochastic action $S(X)$, one can derive the Euler-Lagrange equations. In Appendix F.1, this is done in the Stratonovich formulation, which yields the Stratonovich-Euler-Lagrange equations

$$d_\circ \frac{\partial L^\circ}{\partial v_\circ^i} = \frac{\partial L^\circ}{\partial x^i}\, dt. \tag{3.42}$$

For the Lagrangian (3.37), this becomes

$$d_\circ \left(m\, \delta_{ij} V_\circ^j + q\, A_i\right) = \left(q\, \partial_i A_j\, V_\circ^j - \partial_i \mathfrak{U}\right) dt. \tag{3.43}$$

Hence, using Eq. (3.28), one finds

$$d_\circ X^i = V_\circ^i\, dt,$$
$$m\, \delta_{ij}\, d_\circ V_\circ^j = q\, F_{ij}\, V_\circ\, dt - q\, \partial_t A_i\, dt - \partial_i \mathfrak{U}\, dt. \tag{3.44}$$

This is a set of stochastic differential equations in the sense of Stratonovich and is the stochastic equivalent of the classical equation (2.8).

Alternatively, one can work in the Itô formulation. We minimize the action for the Itô Lagrangians in Appendix F.2, which yields the Itô-Euler-Lagrange equations

$$d_{\pm} \frac{\partial L^{\pm}}{\partial v_{\pm}^i} = \frac{\partial L^{\pm}}{\partial x^i}\, dt. \tag{3.45}$$

For the Lagrangian (3.40), the forward equations become

$$d_+ \left(m\, \delta_{ij} V_+^j + q\, A_i \right) = \left(q\, \partial_i A_j\, V_+^j + \frac{q}{2}\, \partial_i \partial_j A_k\, V_2^{jk} - \partial_i \mathfrak{U} \right) dt \,. \tag{3.46}$$

Hence, using Eqs. (3.29) and (3.31), one finds

$$d_+ X^i = V_+^i\, dt \,,$$

$$d[X^i, X^j] = V_2^{ij}\, dt \,,$$

$$m\, \delta_{ij}\, d_+ V_+^j = q\, F_{ij}\, V_+^j\, dt + \frac{q}{2}\, \partial_k F_{ij}\, V_2^{jk}\, dt - q\, \partial_t A_i\, dt - \partial_i \mathfrak{U} dt \,. \tag{3.47}$$

This is a set of stochastic differential equations in the sense of Itô, and is equivalent to the Stratonovich equation (3.44).

Finally, by a similar calculation, one can derive the equivalent set of backward equations. These are given by

$$d_- X^i = V_-^i\, dt \,,$$

$$d[X^i, X^j] = V_2^{ij}\, dt \,,$$

$$m\, \delta_{ij}\, d_- V_-^j = q\, F_{ij}\, V_-^j\, dt - \frac{q}{2}\, \partial_k F_{ij}\, V_2^{jk}\, dt - q\, \partial_t A_i\, dt - \partial_i \mathfrak{U} dt \,. \tag{3.48}$$

3.5 Boundary Conditions

The stochastic Euler-Lagrange equations, that were derived in previous section, admit solutions. However, uniqueness of the solutions can only be achieved, if appropriate initial conditions are specified. Such initial conditions can be obtained by measuring the particle at an initial time t_0, but the measurement process is different for deterministic and stochastic theories.

In a deterministic theory, one can in theory perform a measurement of both the position X_{t_0} and the velocity V_{t_0} with an infinite precision. However, in practice, one will always have a finite uncertainty on such measurements, since any measurement device has a finite precision. Moreover, a measurement requires an interaction between the measurement device and the particle, which alters the state of the particle, and thus induces an additional uncertainty.

In a stochastic theory, the position process is almost surely not differentiable. Therefore, one cannot provide an initial condition for the process V_t. Nevertheless, one can still obtain unique solutions to the stochastic equations of motion.[13] The

---

[13] Uniqueness refers to the uniqueness of the stochastic process $X : \mathcal{T} \times (\Omega, \Sigma, \mathbb{P}) \to (\mathcal{M}, \mathcal{B}(\mathcal{M}))$. When this stochastic process is evolved in time, it will follow one of all possible sample paths $X(\cdot, \omega) : \mathcal{T} \to \mathcal{M}$, where $\omega \in \Omega$ is chosen according to the probability measure $\mathbb{P}$. Uniqueness of the process does not imply uniqueness of the sample path.

reason for this is that the initial condition that must be seeded into the stochastic differential equation is an initial probability measure $\mu_{X_{t_0}}$. This measure can be obtained from a probability density $\rho(X_{t_0})$, using that $d\mu = \rho d^d x$, which itself can be obtained by measuring the moments $\mathbb{E}[X_{t_0}^n]$ for all $n \in \mathbb{N}$.

The moments $\mathbb{E}[X_{t_0}^n]$ can in theory be measured up to an infinite precision, but in practice there will always be a finite experimental uncertainty, as is the case in deterministic theories. In principle, one could also measure the moments $\mathbb{E}[V_{t_0}^n]$ of the velocity process, which would provide the initial measure for the velocity process $\mu_{V_{t_0}}$. However, this would be in contradiction with the the non-differentiability of X, which implied that initial measure for the process $\mu_{V_{t_0}}$ does not exist.

The resolution of this paradox is that the moments $\mathbb{E}[X_t^n]$ and $\mathbb{E}[V_t^n]$ can be measured at any time t, but the moments of X and of V cannot be measured simultaneously. Hence, if the measure $\mu_{X_{t_0}}$ is constructed, the measure $\mu_{V_{t_0}}$ does not exist and vice versa. This feature leads to a theoretical uncertainty, which is inherently different from the classical experimental uncertainties that we discussed earlier. In the next section, we show that this theoretical uncertainty, that can be formulated as an uncertainty principle, is a generic feature of stochastic theories.

3.6 The Momentum Process

In the previous sections, we have studied a stochastic process $X : \mathcal{T} \times (\Omega, \Sigma, \mathbb{P}) \to (\mathbb{R}^d, \mathcal{B}(\mathbb{R}^d))$ describing the position of a particle with mass m charged under a scalar potential $\mathfrak{U}$ and a vector potential A_i with charge q. We will now construct a dual process P, which we call the momentum process.

As reviewed in Appendix B, the stochastic process X is a family of random variables $\{X_t \mid t \in \mathcal{T}\}$, and, for every $t \in \mathcal{T}$, $X_t \in L_t^2(\Omega, \Sigma, \mathbb{P})$. This allows to consider elements of the dual space, $\tilde{P}_t \in L_t^2(\Omega, \Sigma, \mathbb{P})^*$, which are maps $\tilde{P}_t : L_t^2(\Omega, \Sigma, \mathbb{P}) \to \mathbb{R}$. Since L^2-spaces are self-dual, there exists a family of isomorphisms $\phi_t : (L_t^2)^* \to L_t^2$. Hence, we can define a momentum process $\{P_t \mid t \in \mathcal{T}\}$, where $P_t = \phi_t(\tilde{P}_t) : (\Omega, \Sigma, \mathbb{P}) \to (\mathbb{R}^d, \mathcal{B}(\mathbb{R}^d))$ is a momentum random variable for every $t \in \mathcal{T}$.

We can consider the process (X, P) on the cotangent bundle $T^*\mathbb{R}^d \cong \mathbb{R}^{2d}$. When studying this process, we would like to condition (X, P) on its previous or future states. For this, we require the notion of a filtration, cf. Appendix B.2. However, the processes X and P are mutually incompatible in the sense that they are not adapted to each others filtration.

We already encountered this incompatibility when studying the process (X, V) on the tangent bundle $T\mathbb{R}^d \cong \mathbb{R}^{2d}$. There, we noted that the process V is not differentiable, which necessitated the introduction of Itô processes $V_\pm$ and a Stratonovich process $V_\circ = \frac{1}{2}(V_+ + V_-)$. In a position representation,[14] these velocity processes are integrable, such that

$$\mathbb{E}[V_t \,|\, X_t] = v(X_t, t) \qquad \forall t \in \mathcal{T}, \tag{3.49}$$

but not square integrable, as

$$\mathbb{E}\left[|V_t|^2 \,|\, X_t\right] = \infty \qquad \forall t \in \mathcal{T}. \tag{3.50}$$

The same reasoning can be applied to the momentum processes: one can introduce Itô processes $P_\pm$ and a Stratonovich process $P_\circ = \frac{1}{2}(P_+ + P_-)$, and the expected value for these processes is given by

$$\mathbb{E}[P_t \,|\, X_t] = p(X_t, t) \qquad \forall t \in \mathcal{T}, \tag{3.51}$$

but

$$\mathbb{E}\left[|P_t|^2 \,|\, X_t\right] = \infty \qquad \forall t \in \mathcal{T}. \tag{3.52}$$

Alternatively, one could work in a momentum representation,[15] and introduce Itô position processes $X_\pm$ and a Stratonovich process $X_\circ = \frac{1}{2}(X_+ + X_-)$. The expected value for these processes is given by

$$\mathbb{E}[X_t \,|\, P_t] = x(P_t, t) \qquad \forall t \in \mathcal{T}, \tag{3.53}$$

but

$$\mathbb{E}\left[|X_t|^2 \,|\, P_t\right] = \infty \qquad \forall t \in \mathcal{T}. \tag{3.54}$$

We can estimate the divergences appearing in the second moment of P in the position representation and in the second moment of X in the momentum representation.[16]

Suppose that we are given the state M_0 of a Wiener process at time t_0. We can calculate its quadratic variation for any time $t \geq t_0$, and find

$$[M^a, M^b]_t = \int_{t_0}^t d[M^a, M^b]_s = \frac{\alpha}{m}(t - t_0)\,\delta^{ab}. \tag{3.55}$$

This expression yields the covariance matrix of the process M, given by

---

[14] I.e. with respect to the to the natural filtration $\mathcal{F}^X$ of X.

[15] I.e. with respect to the to the natural filtration $\mathcal{F}^P$ of P.

[16] In this book, we work in a position representation. Therefore, we only discuss the divergence appearing in (3.52), but a similar reasoning can be applied to (3.54).

$$\mathrm{Cov}_{t_0}(M_t^a, M_t^b) = \mathbb{E}[M_t^a M_t^b \mid M_0] - \mathbb{E}[M_t^a \mid M_0]\,\mathbb{E}[M_t^b \mid M_0]$$

$$= \frac{\alpha}{m}\,(t - t_0)\,\delta^{ab}\,. \tag{3.56}$$

The velocity process $\dot{M}$ is not adapted to M. Therefore, the quadratic variation of $\dot{M}$ is not well defined in the position representation. We can, however, write down a formal expression by inverting the time parameter. This yields

$$[\dot{M}^a, \dot{M}^b]_t = \frac{\alpha}{m}\,(t - t_0)^{-1}\,\delta^{ab}\,. \tag{3.57}$$

As expected, this term contains a divergence at t_0, but we have now identified the order of this divergence. Furthermore, the covariance matrix follows from this expression and is given by

$$\mathrm{Cov}_{t_0}(\dot{M}_t^a, \dot{M}_t^b) = \mathbb{E}[\dot{M}_t^a \dot{M}_t^b \mid M_0] - \mathbb{E}[\dot{M}_t^a \mid M_0]\,\mathbb{E}[\dot{M}_t^b \mid M_0]$$

$$= \frac{\alpha}{m}\,(t - t_0)^{-1}\,\delta^{ab}\,. \tag{3.58}$$

Then, for any time t, the product of the covariance matrices is given by

$$\mathrm{Cov}(\dot{M}^a, \dot{M}^b)\,\mathrm{Cov}(M^c, M^d) = \frac{\alpha^2}{m^2}\,\delta^{ab}\,\delta^{cd}\,. \tag{3.59}$$

This defines an uncertainty principle, as it provides a minimal theoretical uncertainty on the joint process $(M, \dot{M})$, which takes values in $T_\circ\mathbb{R}^d$. This derivation can be made more rigorous using the isomorphisms $\phi_t : (L_t^2)^* \to L_t^2$. These induce a map from the characteristic functional φ_{M_t} to the probability density $\rho_{\dot{M}_t}$ implying that ρ_{M_t} and $\rho_{\dot{M}_t}$ are related by a Fourier transform.

We will now consider the quadratic covariation of the processes M and $\dot{M}$. As is the case for the quadratic variation of $\dot{M}$ in the position representation, this covariation is not well defined, but we can give a formal expression. For this, we first write the quadratic variation of $\dot{M}$ in its differential form:

$$d[\dot{M}^a, \dot{M}^b]_t = \frac{\alpha}{m}\,\delta^{ab}\,d(t - t_0)^{-1}$$

$$= -\frac{\alpha}{m}\,(t - t_0)^{-2}\,\delta^{ab}\,dt\,. \tag{3.60}$$

In a similar fashion, we find that the covariation of $\dot{M}$ and M is given by

$$d[\dot{M}^a, M^b]_t = \frac{\alpha}{m}\,\delta^{ab}\,(t - t_0)\,d(t - t_0)^{-1}$$

$$= -\frac{\alpha}{m}\,(t - t_0)^{-1}\,\delta^{ab}\,dt \tag{3.61}$$

and

$$d[M^a, \dot{M}^b]_t = \frac{\alpha}{m} \delta^{ab} (t - t_0)^{-1} d(t - t_0)$$

$$= +\frac{\alpha}{m} (t - t_0)^{-1} \delta^{ab} dt . \tag{3.62}$$

Here, we note that the covariation $[M, \dot{M}]$ is not symmetric, which reflects the mutual incompatibility of the processes M and $\dot{M}$.[17] Furthermore, the sign of these expressions depends on a time ordering convention, such that the opposite time ordering leads to opposite signs.

Equations (3.61) and (3.62) suggest a non-commutativity of M and $\dot{M}$. This can be made explicit by interpreting the commutator as the difference of a time ordered product, such that

$$[M^a, \dot{M}^b] = \lim_{dt \to 0} M^a_{t+dt} \dot{M}^b_t - \dot{M}^b_{t+dt} M^a_t$$

$$= \lim_{dt \to 0} \frac{M^a_{t+dt} (M^b_{t+dt} - M^b_t) - (M^b_{t+dt} - M^b_t) M^a_t}{dt}$$

$$= \lim_{dt \to 0} \frac{(M^a_{t+dt} - M^a_t) (M^b_{t+dt} - M^b_t)}{dt}$$

$$= \lim_{dt \to 0} \frac{d[M^a, M^b]_t}{dt}$$

$$= \frac{\alpha}{m} \delta^{ab} . \tag{3.63}$$

The expressions (3.61) and (3.62) for the covariation of M and $\dot{M}$ can be used to give meaning to the covariation $[X, V_\circ]$ that is encountered in the divergent part of the Itô Lagrangian (3.41). This covariation can be derived by identifying $P = P_\circ$ and using that the momentum P of $X = C + M$ is given by

$$P_i = \frac{\partial L}{\partial v^i} = m \, \delta_{ij} V^j + q \, A_i . \tag{3.64}$$

Moreover, the commutator (3.63) implies a commutation relation for the position and momentum, which is given by

$$[X^i, P_j] = m \, [X^i, \delta_{jk} V^k + q \, A_j(X)]$$

$$= m \, \delta_{jk} [X^i, V^k]_c + m \, q \, [X^i, A_j(X)]$$

$$= m \, \delta_{jk} \delta^i_a \delta^k_b [M^a, \dot{M}^b] + m \, q \, \partial_k A_j(X) \, \delta^i_a \delta^k_b [M^a, M^b],$$

$$= \alpha \, \delta^i_j . \tag{3.65}$$

---

[17] The limit in the definition (B.11) of the quadratic covariation does not converge to a unique value for $[M, \dot{M}]$.

3.7 Stochastic Hamilton-Jacobi Equations

In this section, we construct the stochastic equations of motion in the Hamilton-Jacobi formalism. We define Hamilton's principal function for L^+ as

$$S^+(x, t) = S^+(x, t; x_f, t_f)$$
$$= -\mathbb{E}\left[\int_t^{t_f} L^+(X_s, V_{+,s}, V_{2,s}, s)\, ds \,\Big|\, X_t = x, X_{t_f} = x_f\right], \quad (3.66)$$

where (X_s, V_s) is a solution of the stochastic Itô-Euler-Lagrange equations passing through (x, t) and (x_f, t_f). Moreover, for L^-, we define the principal function as

$$S^-(x, t) = S^-(x, t; x_0, t_0)$$
$$= \mathbb{E}\left[\int_{t_0}^t L^-(X_s, V_{-,s}, V_{2,s}, s)\, ds \,\Big|\, X_t = x, X_{t_0} = x_0\right], \quad (3.67)$$

where (X_s, V_s) is a solution of the Itô-Euler-Lagrange equations passing through (x_0, t_0) and (x, t).

Using Hamilton's principal function, one can derive a set of stochastic Hamilton-Jacobi equations, which is done in Appendix F.3. We find

$$\begin{cases} \frac{\partial}{\partial x^i} S^\pm(x, t) &= p_i^\pm \\ \frac{\partial}{\partial t} S^\pm(x, t) &= -H_0^\pm(x, p^\pm, \partial p^\pm, t) \end{cases} \quad (3.68)$$

with the momentum given by

$$p_i^\pm = \frac{\partial L_0^\pm(x, v_\pm, v_2, t)}{\partial v_\pm^i} \quad (3.69)$$

and the Hamiltonian by

$$H_0^\pm(x, p^\pm, \partial p^\pm, t) = p_i^\pm v_\pm^i \pm \frac{1}{2} \partial_j p_i^\pm v_2^{ij} - L_0^\pm(x, v_\pm, v_2, t). \quad (3.70)$$

For the Itô Lagrangian (3.40), this yields

$$\begin{cases} \frac{\partial S^\pm}{\partial x^i} = m\, \delta_{ij} v_\pm^j + q\, A_i, \\ \frac{\partial S^\pm}{\partial t} = -\frac{m}{2} \delta_{ij} v_\pm^i v_\pm^j \mp \frac{m}{2} \delta_{ij} v_2^{ik} \partial_k v_\pm^j - \mathfrak{U}. \end{cases} \quad (3.71)$$

In addition, we find an integral constraint for the velocity field $v_\pm(X_t, t)$ given by

$$\oint \left(p_i^\pm v_\pm^i \pm \frac{1}{2} v_2^{ij} \partial_j p_i^\pm \right) dt = \pm \mathbb{E}\left[\oint L_\infty(X_s, V_{o,s})\, ds \,\Big|\, X_t\right]. \quad (3.72)$$

For the divergent Lagrangian (3.41), the right hand side yields

$$\mathbb{E}\left[\oint_\gamma L_\infty \, ds \,\Big|\, X_t\right] = \mathbb{E}\left[\oint_\gamma \frac{m}{2} \delta_{ij} \, d[X^i, V_\circ^j]_s \,\Big|\, X_t\right]$$

$$= \mathbb{E}\left[\oint_\gamma \frac{1}{2}\left(d[X^i, P_{\circ,i}]_s - q \, \partial_j A_i \, d[X^i, X^j]_s\right)\Big|\, X_t\right]$$

$$= \frac{\alpha}{2} \oint_\gamma \left(\frac{\delta_i^i}{s-t} - \frac{q}{m} \partial_i A^i\right) ds$$

$$= \alpha \pi \, \mathrm{i} \, k_i^i \,, \tag{3.73}$$

where $k_j^i \in \mathbb{Z}^{d \times d}$ is a matrix of winding numbers counting the number of times the loop γ winds around the pole at $s = t$.

As was done in the classical theory, we can rewrite the Eq. (3.71) into a form that is equivalent to the Itô-Euler-Lagrange equations (3.47) and (3.48). This can be achieved by taking a spatial derivative of the second equation and plugging in the first equation. This yields

$$m \, \delta_{ij} \frac{\partial}{\partial t} v_\pm^j + q \frac{\partial}{\partial t} A_i = -m \, \delta_{jk} v_\pm^k \partial_i v_\pm^j \mp \frac{m}{2} \delta_{jk} v_2^{kl} \partial_l \partial_i v_\pm^j - \partial_i \mathfrak{U} \,. \tag{3.74}$$

Then, using Eq. (2.18), i.e.

$$m \, \delta_{jk} \, \partial_i v_\pm^j = m \, \delta_{ij} \, \partial_k v_\pm^j - q \, F_{ik} \,,$$

we find

$$\left[m \, \delta_{ij} \left(\partial_t + v_\pm^k \partial_k \pm \frac{1}{2} v_2^{kl} \partial_l \partial_k\right) - q \, F_{ij}\right] v_\pm^j = \pm \frac{q}{2} v_2^{jk} \partial_k F_{ij} - q \, \partial_t A_i - \partial_i \mathfrak{U} \,. \tag{3.75}$$

When supplemented with

$$d_\pm X_t^i = v_\pm^i(X_t, t) \, dt + \delta_a^i \, dM_t^a \,, \tag{3.76}$$

$$d[M^a, M^b]_t = \frac{\alpha}{m} \delta^{ab} \, dt \,, \tag{3.77}$$

this set of equations is equivalent to the Itô-Euler-Lagrange equations (3.47) and (3.48) with the process $V_{\pm,t}$ replaced by the field $v_\pm(X_t, t)$.

3.8 Diffusion Equations

As anticipated in Chap. 2, one can derive a partial differential equation for Hamilton's principal functions $S^\pm(X_t, t)$. Combining the stochastic Hamilton-Jacobi equations (3.71) and plugging in the expression (3.33) for v_2 yields

$$-2m \frac{\partial S^\pm}{\partial t} = \partial_i S^\pm \partial^i S^\pm \pm \alpha \partial_i \partial^i S^\pm - 2q A^i \partial_i S^\pm \mp \alpha q \partial_i A^i + q^2 A_i A^i + 2m \mathfrak{U}.$$
(3.78)

This is the generalization of Eq. (2.21) to a stochastic theory.

We can define the wave functions

$$\Psi_\pm(x, t) := \exp\left(\pm \frac{S^\pm(x, t)}{\alpha}\right).$$
(3.79)

Then, Eq. (3.78) implies that $\Psi_+(x, t)$ is a solution of the time reversed heat equation

$$-\alpha \frac{\partial}{\partial t} \Psi = \left[\frac{\delta^{ij}}{2m}\left(\alpha \frac{\partial}{\partial x^i} - q A_i\right)\left(\alpha \frac{\partial}{\partial x^j} - q A_j\right) + \mathfrak{U}\right]\Psi$$
(3.80)

subjected to a terminal condition $\Psi_+(x, t_f) = u_f(x)$. Moreover, $\Psi_-(x, t)$ is a solution of the heat equation

$$\alpha \frac{\partial}{\partial t} \Psi = \left[\frac{\delta^{ij}}{2m}\left(\alpha \frac{\partial}{\partial x^i} + q A_i\right)\left(\alpha \frac{\partial}{\partial x^j} + q A_j\right) + \mathfrak{U}\right]\Psi$$
(3.81)

subjected to an initial condition $\Psi_-(x, t_0) = u_0(x)$. Thus, we can associate solutions of the heat equation to solutions of the stochastic Hamilton-Jacobi and Euler-Lagrange equations.

Conversely, we would like to associate a stochastic process X to any solution of the diffusion equations (3.80) and (3.81). Here, we encounter a small caveat, as Hamilton's principal function is multi-valued: due to the integral constraint (3.72), Hamilton's principal function defines an equivalence class $[S^\pm]$ under the equivalence relation

$$\tilde{S}^\pm \sim S^\pm \quad \text{if} \quad \tilde{S}^\pm = S^\pm + \alpha \pi i \sum_{i=1}^{d} k_i \quad \text{with } k_i \in \mathbb{Z}.$$
(3.82)

This implies that the wave functions (3.79) define equivalence classes $[\Psi_\pm]$ with equivalence relations

$$\begin{cases} \tilde{\Psi}_+ \sim \Psi_+ & \text{if} \quad \tilde{\Psi}_+ = \pm \Psi_+ \,, \\ \tilde{\Psi}_- \sim \Psi_- & \text{if} \quad \tilde{\Psi}_- = \pm \Psi_- \,, \end{cases} \tag{3.83}$$

which must be taken into account when formulating the converse statement.

Hence, we find that, for any solution Ψ_+ of Eq. (3.80) and Ψ_- of Eq. (3.81) modulo the equivalence relation (3.83), one can construct velocity fields

$$v_\pm^i = \frac{\delta^{ij}}{m} \left(\pm \alpha \, \partial_j \ln \Psi_\pm - q \, A_j \right), \tag{3.84}$$

such that solutions X of the Itô equation

$$d_\pm X_t^i = v_\pm^i(X_t, t) \, dt + \delta_a^i \, dM_t^a \tag{3.85}$$

$$d[M^a, M^b]_t = \frac{\alpha}{m} \, \delta^{ab} \, dt \tag{3.86}$$

minimize the stochastic action $S(X)$ associated to Hamilton's principal function

$$S_\pm = \alpha \, \ln \Psi_\pm \,. \tag{3.87}$$

We conclude that there is a correspondence between solutions of the diffusion equations (3.80) and (3.81) with $\alpha > 0$ modulo the equivalence relation (3.83) and semi-martingale processes X that minimize the stochastic action $S(X)$ and satisfy the structure relation (3.32). This correspondence is similar to the one established by the Feynman-Kac theorem [24], where the time reversed heat equation (3.80) is the Kolmogorov backward equation for the process X, when evolved forward in time, and the heat equation (3.81) is the Kolmogorov backward equation for the process X, when evolved backward in time.

We conclude this section by pointing out that in earlier formulations of stochastic mechanics, the divergent part of the Lagrangian is discarded, as it does not contribute to the equations of motion for the process X. However, when this is done, one does not have an equivalence between solutions of the of the diffusion equations and solutions of the Hamilton-Jacobi equations.

Indeed, for any solution Ψ of the diffusion equations, one can construct an equivalent solution $\Psi \exp(2\pi i \sum_{i=1}^d k_i)$, which imposes an equivalence relation on Hamilton's principal functions $S^\pm$. However, when the divergent part of the Lagrangian is ignored, such an equivalence relation is not present in the theory and must be imposed by hand.

This issue is known as Wallstrom's criticism of stochastic mechanics [99, 100]. Here, we have resolved this criticism, as we have shown that such an equivalence relation follows from the integral condition (3.72), which is part of the theory, when the divergent part of the Lagrangian is properly taken into account.

Chapter 4
Complex Stochastic Dynamics on $\mathbb{R}^d$

Abstract This chapter extends the stochastic theory from Chap. 3 to complex diffusions. It introduces a complex superposition of two Wiener processes defined on the reference frames, and studies their real projection on the configuration space. Then, it shows that this process generates the dynamics of a quantum theory in the limit where the Wiener processes are maximally correlated, and a Brownian theory when the processes are uncorrelated.

In the previous chapter, we studied stochastic processes on $\mathbb{R}^d$. More precisely, we studied semi-martingale processes $X = C + M$ that can be decomposed in a deterministic trajectory C and a noise term M. Then, we imposed the process M to be a Wiener process using the structure relation (3.17), and we used the stationary action principle to derive equations of motion for the process X. Finally, using the equations of motion, we derived a correspondence between solutions of real diffusion equations and semi-martingale processes that minimize the stochastic action.

Given the fact that the Schrödinger equation is a complex diffusion equation, one might wonder whether our analysis can be generalized, such that a correspondence between a generalized stochastic process and complex diffusion equations is obtained. In this chapter, we show that this can be done, when the martingale M is complexified. We will do this by complexifying the fibers to $F = \mathbb{C}^d$, while keeping the base space $\mathcal{M} = \mathbb{R}^d$ real.

In order to incorporate complex martingale processes, while keeping the process $X \in \mathbb{R}^d$ real, we will consider a semi-martingale that can be decomposed as

$$\begin{aligned} X_t^i &= C_{x,t}^i + \delta_a^i \operatorname{Re}\left[M_t^a\right] \\ &= C_{x,t}^i + \delta_a^i M_{x,t}^a, \end{aligned} \qquad (4.1)$$

where $C_{x,t}$ is a real valued continuous trajectory with finite variation on $\mathbb{R}^d$ and M is a complex valued martingale on $\mathbb{C}^d$, i.e.

$$M_t = M_{x,t} + \mathrm{i}\, M_{y,t}, \qquad (4.2)$$

where M_x, M_y are real d-dimensional martingales.

© The Author(s), under exclusive license to Springer Nature Switzerland AG 2023
F. Kuipers, *Stochastic Mechanics*, SpringerBriefs in Physics,
https://doi.org/10.1007/978-3-031-31448-3_4

We will make the assumption that M_x and M_y are Lévy processes[1] and therefore also Markov processes. As before, we will further specify M using a structure relation. In particular, we will consider

$$m\,d[M^a, M^b]_t = \alpha\,\delta^{ab}\,dt$$
$$= |\alpha|\,e^{i\phi}\,\delta^{ab}\,dt \tag{4.3}$$

with $\alpha \in \mathbb{C}$. Hence, compared to the structure relation (3.17) from previous chapter, we have added the complex factor $e^{i\phi}$. As we are studying a complex process, we must also specify the structure relations for the conjugated process $\overline{M}$. A general complex structure relation is of the form

$$d\begin{pmatrix} [M^a, M^b]_t & [M^a, \overline{M}^b]_t \\ [\overline{M}^a, M^b]_t & [\overline{M}^a, \overline{M}^b]_t \end{pmatrix} = \frac{\delta^{ab}}{m}\begin{pmatrix} \alpha & |\alpha| + \gamma \\ |\alpha| + \gamma & \overline{\alpha} \end{pmatrix} dt\,. \tag{4.4}$$

with[2] $\alpha \in \mathbb{C}$ and $\gamma \in [0, \infty)$. Such a general process requires new degrees of freedom, as it defines a diffusion in $\mathbb{C}^d$ instead of $\mathbb{R}^d$. We can resolve this issue by setting $\gamma = 0$, such that

$$d\begin{pmatrix} [M^a, M^b]_t & [M^a, \overline{M}^b]_t \\ [\overline{M}^a, M^b]_t & [\overline{M}^a, \overline{M}^b]_t \end{pmatrix} = \frac{\delta^{ab}}{m}\begin{pmatrix} \alpha & |\alpha| \\ |\alpha| & \overline{\alpha} \end{pmatrix} dt\,. \tag{4.5}$$

For this choice, the structure relations are degenerate, which implies that the complex process $M \in \mathbb{C}^n$ has only d degrees of freedom instead of $2d$, such that M is restricted to the subspace $e^{\frac{i\phi}{2}} \times \mathbb{R}^d \subset \mathbb{C}^d$. Hence, the structure relation (4.5) characterizes a real Wiener process that is rotated in the complex plane over an angle $\frac{\phi}{2}$.

We can also define an auxiliary process that can be decomposed as

$$Y_t^i = C_{y,t}^i + \delta_a^i\,\mathrm{Im}\left[M_t^a\right]$$
$$= C_{y,t}^i + \delta_a^i\,M_{y,t}^a\,, \tag{4.6}$$

where $C_{y,t}$ is another $\mathbb{R}^d$-valued continuous trajectory with finite variation. In addition, we write

$$Z_t = X_t + i\,Y_t \tag{4.7}$$

and

$$C_t = C_{x,t} + i\,C_{y,t}\,. \tag{4.8}$$

---

[1] Note that M_t is not a complex Lévy process, as we do not assume that M_x and M_y are independent.

[2] These are necessary and sufficient conditions for the existence of the complex martingale M. For $\alpha = 0$, the process is the complex Wiener process, which is an example of a conformal martingale.

Using Eq. (4.5), we find the quadratic covariation of the processes X and Y, which is given by

$$d \begin{pmatrix} [X^a, X^b]_t & [X^a, Y^b]_t \\ [Y^a, X^b]_t & [Y^a, Y^b]_t \end{pmatrix} = \frac{|\alpha|}{2m} \delta^{ab} \begin{pmatrix} 1 + \cos\phi & \sin\phi \\ \sin\phi & 1 - \cos\phi \end{pmatrix} dt . \tag{4.9}$$

As in previous chapter, one can derive the equations of motion using a Stratonovich formulation, a forward Itô formulation or a backward Itô formulation. In order to do so, we must introduce the velocity fields. These are now given by the complex fields

$$w^i(X_t, t) = v^i(X_t, t) + i\, u^i(X_t, t). \tag{4.10}$$

Since the stochastic motion is not differentiable, we obtain a forward and a backward velocity field

$$w^i_+(X_t, t) = \lim_{h \to 0} \mathbb{E}\left[\frac{Z^i_{t+dt} - Z^i_t}{dt} \,\Big|\, X_t \right], \tag{4.11}$$

$$w^i_-(X_t, t) = \lim_{h \to 0} \mathbb{E}\left[\frac{Z^i_t - Z^i_{t-dt}}{dt} \,\Big|\, X_t \right]. \tag{4.12}$$

In addition, we define a Stratonovich velocity field

$$\begin{aligned} w^i_\circ(X_t, t) &= \lim_{h \to 0} \mathbb{E}\left[\frac{Z^i_{t+dt} - Z^i_{t-dt}}{2\,dt} \,\Big|\, X_t \right] \\ &= \frac{1}{2}\left[w^i_+(X_t, t) + w^i_-(X_t, t) \right] \end{aligned} \tag{4.13}$$

and a second order velocity field

$$w^{ij}_2(X_t, t) = \lim_{h \to 0} \mathbb{E}\left[\frac{[Z^i_{t+dt} - Z^i_t][Z^j_{t+dt} - Z^j_t]}{dt} \,\Big|\, X_t \right], \tag{4.14}$$

which can be decomposed as

$$w^{ij}_2(X_t, t) = v^{ij}_2(X_t, t) + i\, u^{ij}_2(X_t, t), \tag{4.15}$$

and is fixed by the structure relation (4.3), such that

$$w^{ij}_2(X_t, t) = \frac{\alpha}{m} \delta^{ij} . \tag{4.16}$$

4.1   Stochastic Action

As in previous chapter, we must specify the phase space in order to define a Lagrangian. This phase space can be obtained by complexifying the tangent bundles from previous chapter. The Stratonovich bundle becomes $T_\circ^{\mathbb{C}} \mathbb{R}^d \cong \mathbb{R}^d \times \mathbb{C}^d$ and the Itô bundles become $T_\pm^{\mathbb{C}} \mathbb{R}^d \cong \mathbb{R}^d \times \mathbb{C}^{\frac{d(d+3)}{2}}$. This allows to study processes $(X_t, W_{\circ,t})$ on $T_\circ^{\mathbb{C}} \mathbb{R}^d$ and $(X_t, W_{\pm,t}, \pm W_{2,t})$ on $T_\pm^{\mathbb{C}} \mathbb{R}^d$, using the relations[3]

$$W_{\circ,t}^i \, dt = d_\circ Z_t^i \,, \tag{4.17}$$

$$W_{+,t}^i \, dt = d_+ Z_t^i \,, \tag{4.18}$$

$$W_{-,t}^i \, dt = d_- Z_t^i \,, \tag{4.19}$$

$$W_{2,t}^{ij} \, dt = d[Z^i, Z^j]_t \,. \tag{4.20}$$

We construct Lagrangians on these complexified tangent bundles by replacing the real velocity fields with complex fields. The Stratonovich Lagrangian is thus given by

$$L^\circ(x, w_\circ, t) = \frac{m}{2} \, \delta_{ij} \, w_\circ^i w_\circ^j + q \, A_i(x, t) \, w_\circ^i - \mathfrak{U}(x, t) \,, \tag{4.21}$$

and the Itô Lagrangians by

$$L^\pm(x, w_\pm, w_2, t) = L_0^\pm(x, w_\pm, w_2, t) \pm L_\infty(x, w_\circ) \tag{4.22}$$

with finite part

$$L_0^\pm(x, w_\pm, w_2, t) = \frac{m}{2} \, \delta_{ij} \, w_\pm^i w_\pm^j + q \, A_i \, w_\pm^i \pm \frac{q}{2} \, \partial_j A_i \, w_2^{ij} - \mathfrak{U} \tag{4.23}$$

and a divergent part that satisfies the integral condition

$$\mathbb{E}\left[\int L_\infty(x, w_\circ) \, dt \right] = \mathbb{E}\left[\int \frac{m}{2} \, \delta_{ij} \, d[z^i, w_\circ^j] \right]. \tag{4.24}$$

4.2   Equations of Motion

All results from the real diffusion theory can be generalized to the complex case. The Stratonovich-Euler-Lagrange equations (3.42) for the Lagrangian (4.21) are given by

---

[3] As in previous chapter, these differential expressions are defined by an integral expression similar to the one given in Eq. (3.27).

$$d_o Z^i = W_o^i \, dt \, ,$$

$$m \, \delta_{ij} \, d_o W_o^j = q \, F_{ij} \, W_o^j \, dt - q \, \partial_t A_i \, dt - \partial_i \mathfrak{U} dt \, . \tag{4.25}$$

Moreover, the Itô-Euler-Lagrange equations (3.45) for the Lagrangian (4.22) become

$$d_\pm Z^i = W_\pm^i \, dt \, ,$$

$$d[Z^i, Z^j] = W_2^{ij} \, dt \, ,$$

$$m \, \delta_{ij} \, d_\pm W_\pm^j = q \, F_{ij} \, W_\pm^j \, dt \pm \frac{q}{2} \, \partial_k F_{ij} \, W_2^{jk} \, dt - q \, \partial_t A_i \, dt - \partial_i \mathfrak{U} dt \, . \tag{4.26}$$

Hamilton's principal functions are complex valued functions given by

$$S^+(x, t) = -\mathbb{E}\left[\int_t^{t_f} L^+(X_s, W_{+,s}, W_{2,s}, s) \, ds \, \Big| \, X_t = x, X_{t_f} = x_f \right] ,$$

$$S^-(x, t) = \mathbb{E}\left[\int_{t_0}^t L^-(X_s, W_{-,s}, W_{2,s}, s) \, ds \, \Big| \, X_t = x, X_{t_0} = x_0 \right] , \tag{4.27}$$

and can be used to derive the Hamilton-Jacobi equations. These are given by Eq. (3.68) with complex momenta and velocities. For the Lagrangian (4.22), we then find

$$\left[m \, \delta_{ij} \left(\partial_t + w_\pm^k \partial_k \pm \frac{1}{2} w_2^{kl} \partial_l \partial_k \right) - q \, F_{ij} \right] w_\pm^j = \pm \frac{q}{2} \, w_2^{jk} \partial_k F_{ij} - q \, \partial_t A_i - \partial_i \mathfrak{U} , \tag{4.28}$$

which is equivalent to Eq. (4.26) with W_t replaced by $w(X_t, t)$. In addition, the Hamilton-Jacobi equations provide an integral constraint for the velocity field $w_\pm(X_t, t)$, given by

$$\oint \left(p_i^\pm w_\pm^i \pm \frac{1}{2} w_2^{ij} \partial_j p_i^\pm \right) dt = \pm \alpha \, \pi \, i \, k_i^i \, , \tag{4.29}$$

where $k_j^i \in \mathbb{Z}^{d \times d}$ is a matrix of winding numbers.

Taking into account this integral constraint, the Hamilton-Jacobi equations can be solved for the velocity fields $w_\pm$, and the solution can be plugged into the stochastic differential equation

$$\begin{cases} d_\pm Z_t^i & = w_\pm^i(X_t, t) \, dt + \delta_a^i \, dM_t^a \, , \\ d[M^a, M^b]_t & = \frac{\alpha}{m} \, \delta^{ab} \, dt \, . \end{cases} \tag{4.30}$$

This Itô equation can be solved for Z yielding solutions $X_t = \text{Re}[Z_t]$ and $Y_t = \text{Im}[Z_t]$, which describe respectively the particle and the auxiliary process.

4.3 Boundary Conditions

The complex stochastic Euler-Lagrange equations that were derived in the previous section admit solutions, and these are unique, when appropriate boundary conditions are provided. However, in the complex theory, there are new complex degrees of freedom, whose initial conditions must also be specified.

In the deterministic theory, which can be obtained by taking the limit $\alpha \to 0$, one must provide initial or terminal conditions for the positions X, Y and velocities V, U. As discussed in Sect. 3.5, such conditions for X, V can be obtained by measurement, but there is no obvious measurement for the complex position Y and velocity U. However, the equations of motion are independent of the complex position Y. Therefore, one can obtain unique solutions for (X, V, U) without specifying boundary conditions for Y. Despite this simplification, one must still provide initial or terminal conditions for the complex velocity U or equivalently for the field $u(x, t)$. We will return to this point and provide a physical interpretation of the complex velocity field $u(x, t)$ in Sect. 7.4.

In the stochastic theory, one must provide the initial or terminal probability measures μ_X and μ_Y, but, by a similar reasoning, one only requires the initial or terminal measure μ_X supplemented with initial or terminal conditions for the velocity fields $u_\pm(x, t)$. As in the deterministic case, the fields, $u_\pm$ do not have a physical interpretation yet, but an interpretation will be provided in Sect. 7.4 for the field $u_\circ = \frac{1}{2}(u_+ + u_-)$. This interpretation fixes both u_+ and u_-, since the four velocity fields v_+, v_-, u_+ and u_- are not independent, due to the fact that the martingale M is restricted to $e^{\frac{i\phi}{2}} \times \mathbb{R}^d \subset \mathbb{C}^d$. As a consequence, the velocity process W is non-differentiable along this direction, but differentiable perpendicular to this hyperplane, which imposes the constraint

$$\cos\left(\frac{\phi}{2}\right) u_+^i - \sin\left(\frac{\phi}{2}\right) v_+^i = \cos\left(\frac{\phi}{2}\right) u_-^i - \sin\left(\frac{\phi}{2}\right) v_-^i . \tag{4.31}$$

Therefore, there are only three independent velocity fields and, if $\phi \neq \pi$, one can take these fields to be v_+, v_- and $u_\circ$.

We point out that $\phi = 0$ for a Brownian motion, which implies that Eq. (4.31) reduces to

$$u_+(x, t) = u_-(x, t) . \tag{4.32}$$

The real Brownian motion from Chap. 3 can then be recovered by adding the initial condition $u_\circ(x, t_0) = 0$, as this ensures that $u_\pm(x, t) = 0$ for all $t \in \mathcal{T}$.

4.4 Diffusion Equations

Combining the stochastic Hamilton-Jacobi equations (3.71), with the real velocities v replaced by complex velocity fields w, and plugging in Eq. (4.16) for w_2 yields a complex partial differential equation

$$-2m \frac{\partial S^\pm}{\partial t} = \partial_i S^\pm \partial^i S^\pm \pm \alpha \partial_i \partial^i S^\pm - 2q A^i \partial_i S^\pm \mp \alpha q \partial_i A^i + q^2 A_i A^i + 2m \mathfrak{U}. \tag{4.33}$$

This equation implies that the wave functions

$$\Psi_\pm(x, t) = \exp\left(\pm \frac{S^\pm(x, t)}{\alpha}\right)$$

satisfy complex diffusion equations. In particular, $\Psi_+(x, t)$ is a solution of the complex diffusion equation

$$-\alpha \frac{\partial}{\partial t} \Psi = \left[\frac{\delta^{ij}}{2m} \left(\alpha \frac{\partial}{\partial x^i} - q A_i\right) \left(\alpha \frac{\partial}{\partial x^j} - q A_j\right) + \mathfrak{U} \right] \Psi, \tag{4.34}$$

which reduces to the heat equation for $\alpha = -1$ and to the Schrödinger equation for $\alpha = -\mathrm{i}$.

Similarly, $\Psi_-(x, t)$ is a solution of the complex diffusion equation

$$\alpha \frac{\partial}{\partial t} \Psi = \left[\frac{\delta^{ij}}{2m} \left(\alpha \frac{\partial}{\partial x^i} + q A_i\right) \left(\alpha \frac{\partial}{\partial x^j} + q A_j\right) + \mathfrak{U} \right] \Psi, \tag{4.35}$$

which reduces to the heat equation for $\alpha = 1$ and to the Schrödinger equation for $\alpha = \mathrm{i}$.

As in previous chapter, this result implies a correspondence between solutions of the diffusion equations (4.34) and (4.35) modulo the equivalence relation (3.83),

$$\begin{cases} \tilde{\Psi}_+ \sim \Psi_+ & \text{if} \quad \tilde{\Psi}_+ = \pm \Psi_+ \\ \tilde{\Psi}_- \sim \Psi_- & \text{if} \quad \tilde{\Psi}_- = \pm \Psi_-, \end{cases}$$

and semi-martingale processes Z that solve the Itô equation

$$d_\pm Z_t^i = w_\pm^i(X_t, t)\, dt + \delta_a^i\, dM_t^a$$
$$d[M^a, M^b]_t = \frac{\alpha}{m} \delta^{ab}\, dt \tag{4.36}$$

with velocity

$$w_\pm^i = \frac{\delta^{ij}}{m} \left(\pm \alpha \partial_j \ln \Psi_\pm - q A_j\right). \tag{4.37}$$

Chapter 5
Relativistic Stochastic Dynamics on $\mathbb{R}^{d,1}$

Abstract This chapter extends the theory of Chap. 4 to relativistic theories by changing the configuration space from the Euclidean space to a Minkowski space and by imposing a stochastic version of the energy-momentum relation. The resulting theory describes both relativistic Brownian motion and relativistic quantum theories in their respective limit.

In previous chapters, we have discussed non-relativistic stochastic dynamics. In this chapter, we will extend this discussion to relativistic theories. In a relativistic theory, the configuration space $\mathbb{R}^d$ is extended to the Minkowski space $\mathbb{R}^{d,1}$ with coordinates $x^\mu = (x^0, x^i) = (c\,t, x^i)$ and $x^i \in \mathbb{R}^d$, and one studies trajectories $\{X_\lambda : \lambda \in \mathcal{T}\}$ parameterized by an affine parameter λ on this Minkowski space.

The dynamics of a relativistic theory is governed by a Lagrangian $L : T\mathbb{R}^{d,1} \times \mathbb{R}^+ \to \mathbb{R}$ defined on the tangent bundle $T\mathbb{R}^{d,1} \cong \mathbb{R}^{2d,2}$. The relativistic Lagrangian corresponding to the non-relativistic Lagrangian (2.1) is given by

$$L(x, v, \varepsilon) = \frac{1}{2\,\varepsilon}\, \eta_{\mu\nu}\, v^\mu v^\nu - \frac{\varepsilon\, m^2}{2} + q\, A_\mu(x)\, v^\mu\,, \tag{5.1}$$

where $\eta_{\mu\nu} = \mathrm{diag}(-1, +1, ..., +1)$ is the Minkowski metric, and $\varepsilon \in (0, \infty)$ is an auxiliary variable that ensures that the action is invariant under affine reparameterizations

$$\lambda \to \tilde{\lambda} = a\,\lambda + b\,,$$

$$\varepsilon \to \tilde{\varepsilon} = \frac{\varepsilon}{a}\,. \tag{5.2}$$

The relativistic Euler-Lagrange equations can be obtained by extremizing the action. For the Lagrangian (5.1), this yields

$$\frac{dX_\lambda^\mu}{d\lambda} = V_\lambda^\mu \,, \tag{5.3}$$

$$\frac{1}{\varepsilon} \, \eta_{\mu\nu} \, \frac{dV_\lambda^\nu}{d\lambda} = q \, F_{\mu\nu}(X_\lambda) \, V_\lambda^\nu \,, \tag{5.4}$$

$$\eta_{\mu\nu} V_\lambda^\mu V_\lambda^\nu = -\varepsilon^2 \, m^2 \, c^2 \,. \tag{5.5}$$

In addition, one must gauge fix the Lagrange multiplier ε with a gauge condition that depends on the mass:

- If $m^2 > 0$, we fix $\varepsilon = |m|^{-1}$, which fixes the affine parameter to be the proper time, i.e. $\lambda = \tau$.
- If $m = 0$, we fix[1] $\varepsilon = c^2 \, E(\lambda)^{-1}$, where E is the energy of the particle and λ is measured in units of time, i.e. $[\lambda] = T$.
- If $m^2 < 0$, we fix $\varepsilon = c^{-1} |m|^{-1}$, which fixes the affine parameter to be the proper length, i.e. $\lambda = s$.

The classical relativistic theory can be generalized to a stochastic theory by considering semi-martingale processes

$$X_\lambda^\mu = C_{x,\lambda}^\mu + \delta_\alpha^\mu \operatorname{Re}[M_\lambda^\alpha] \,, \tag{5.6}$$

where X and C_x are $\mathbb{R}^{d,1}$-valued and M is $\mathbb{C}^{d+1}$-valued. As in previous chapters, we can then fix the stochastic law of X by imposing the structure relation[2]

$$d[M^\alpha, M^\beta]_\lambda = \alpha \, \varepsilon \, \eta^{\alpha\beta} \, d\lambda \tag{5.7}$$

with $\alpha \in \mathbb{C}$, where we use natural units in which $\hbar, c = 1$. Hence, we obtain

$$d[Z^\mu, Z^\nu]_\lambda = \alpha \, \varepsilon \, \eta^{\mu\nu} \, d\lambda + o(d\lambda) \,, \tag{5.8}$$

where $Z = X + i\,Y$ and $Y^\mu = C_y^\mu + \delta_\alpha^\mu \operatorname{Im}[M^\alpha]$ defines an auxiliary process.

---

[1] For $m = 0$, there is no canonical choice that fixes the affine parameter to a physical parameter, as the strict equality $m = 0$ has already reduced the dimension of the phase space. Another common choice in the literature is $\varepsilon = 1$ for which $[\lambda] = T/M$.

[2] For $m = 0$ with gauge choice $\varepsilon = c^2 \, E(\lambda)^{-1}$, M_λ^α is no longer a Lévy process, as it does not have stationary increments. This does not affect the results, as our analysis requires continuity in probability and independent increments, but does not require stationarity of the increments. Moreover, M can be made into a Lévy process by changing the gauge to $\varepsilon = 1$.

5.1 Equations of Motion

Starting from the relativistic classical Lagrangian (5.1), all results from chapter 4 can be generalized to relativistic theories. The Stratonovich Lagrangian is given by

$$L^{\circ}(x, w_{\circ}, \varepsilon) = \frac{1}{2\varepsilon} \eta_{\mu\nu} w_{\circ}^{\mu} w_{\circ}^{\nu} - \frac{\varepsilon m^2}{2} + q A_{\mu}(x) w_{\circ}^{\mu} \tag{5.9}$$

and the Itô Lagrangian by

$$L^{\pm}(x, w_{\pm}, w_2, \varepsilon) = L_0^{\pm}(x, w_{\pm}, w_2, \varepsilon) \pm L_{\infty}(x, w_{\circ}, \varepsilon) \tag{5.10}$$

with finite part

$$L_0^{\pm}(x, w_{\pm}, w_2, \varepsilon) = \frac{1}{2\varepsilon} \eta_{\mu\nu} w_{\pm}^{\mu} w_{\pm}^{\nu} - \frac{\varepsilon m^2}{2} + q A_{\mu} w_{\pm}^{\mu} \pm \frac{q}{2} \partial_{\nu} A_{\mu} w_2^{\mu\nu} \tag{5.11}$$

and a divergent part that is defined by the integral condition

$$\mathbb{E}\left[\int L_{\infty}^{\pm}(x, w_{\circ}, \varepsilon) \, d\lambda\right] = \mathbb{E}\left[\int \frac{1}{2\varepsilon} \eta_{\mu\nu} \, d[z^{\mu}, w_{\circ}^{\nu}]\right]. \tag{5.12}$$

The Stratonovich-Euler-Lagrange equations (3.42) for the Lagrangian (5.9) are

$$d_{\circ} Z^{\mu} = W_{\circ}^{\mu} \, d\lambda, \tag{5.13}$$

$$\eta_{\mu\nu} \, d_{\circ} W_{\circ}^{\nu} = \varepsilon q \, F_{\mu\nu} \, W_{\circ}^{\nu} \, d\lambda, \tag{5.14}$$

$$\mathbb{E}\left[\eta_{\mu\nu} \, W_{\circ}^{\mu} W_{\circ}^{\nu}\right] = -\varepsilon^2 \, m^2. \tag{5.15}$$

Moreover, the Itô-Euler-Lagrange equations (3.45) for the Lagrangian (5.10) are

$$d_{\pm} Z^{\mu} = W_{\pm}^{\mu} \, d\lambda, \tag{5.16}$$

$$d[Z^{\mu}, Z^{\nu}] = W_2^{\mu\nu} \, d\lambda, \tag{5.17}$$

$$\eta_{\mu\nu} \, d_{\pm} W_{\pm}^{\nu} = \varepsilon q \, F_{\mu\nu} \, W_{\pm}^{\nu} \, d\lambda \pm \frac{\varepsilon q}{2} \, \partial_{\rho} F_{\mu\nu} \, W_2^{\nu\rho} \, d\lambda, \tag{5.18}$$

$$\mathbb{E}\left[\eta_{\mu\nu} \, W_{\pm}^{\mu} W_{\pm}^{\nu}\right] = -\varepsilon^2 \, m^2. \tag{5.19}$$

Hamilton's principal functions are complex valued functions given by

$$S^+(x, \varepsilon, \lambda) = -\mathbb{E}\left[\int_{\lambda}^{\lambda_f} L^+(X_{\tau}, W_{+,\tau}, W_{2,\tau}, \mathcal{E}_{\tau}) \, d\tau \,\Big|\, X_{\lambda} = x, X_{\lambda_f} = x_f, \mathcal{E}_{\lambda} = \varepsilon, \mathcal{E}_{\lambda_f} = \varepsilon_f\right]$$

$$S^-(x, \varepsilon, \lambda) = \mathbb{E}\left[\int_{\lambda_0}^{\lambda} L^-(X_{\tau}, W_{-,\tau}, W_{2,\tau}, \mathcal{E}_{\tau}) \, d\tau \,\Big|\, X_{\lambda} = x, X_{\lambda_0} = x_0, \mathcal{E}_{\lambda} = \varepsilon, \mathcal{E}_{\lambda_0} = \varepsilon_0\right]$$

$$\tag{5.20}$$

and can be used to derive the Hamilton-Jacobi equations. These are given by a relativistic version of (3.68):

$$\begin{cases} \frac{\partial}{\partial x^\mu} S^\pm(x, \varepsilon, \lambda) & = p_\mu^\pm, \\ \left(\frac{\partial}{\partial \lambda} + \frac{d\varepsilon(\lambda)}{d\lambda} \frac{\partial}{\partial \varepsilon} \right) S^\pm(x, \varepsilon, \lambda) & = -p_\mu^\pm w_\pm^\mu \mp \frac{1}{2} \partial_\nu p_\mu^\pm w_2^{\mu\nu} + L_0^\pm(x, w_\pm, w_2, \varepsilon), \end{cases}$$
(5.21)

where the momenta $p^\pm$ are complex valued, and w_2 is determined by Eq. (5.8), such that

$$w_2^{\mu\nu}(X_\lambda) = \alpha \varepsilon \eta^{\mu\nu}.$$
(5.22)

Furthermore, the reparameterization invariance of the theory, given in Eq. (5.2), implies the constraint

$$\frac{\partial S^\pm}{\partial \lambda} + \frac{d\varepsilon}{d\lambda} \frac{\partial S^\pm}{\partial \varepsilon} = 0.$$
(5.23)

In addition, the velocity field $w_\pm(X_\lambda)$ satisfies the integral constraint

$$\oint \left(p_\mu^\pm w_\pm^\mu \pm \frac{1}{2} w_2^{\mu\nu} \partial_\nu p_\mu^\pm \right) d\lambda = \pm \alpha \pi i k_\mu^\mu,$$
(5.24)

where $k_\nu^\mu \in \mathbb{Z}^{n \times n}$ is a matrix of winding numbers.

By taking a covariant derivative of the second Hamilton-Jacobi equation, and plugging in the Lagrangian (5.10), one finds

$$\left[\eta_{\mu\nu} \left(w_\pm^\rho \partial_\rho \pm \frac{1}{2} w_2^{\rho\sigma} \partial_\sigma \partial_\rho \right) - \varepsilon q F_{\mu\nu} \right] w_\pm^\nu = \pm \frac{\varepsilon q}{2} w_2^{\nu\rho} \partial_\rho F_{\mu\nu},$$
(5.25)

which is equivalent to Eq. (5.18) with W replaced by $w(X)$. Taking into account the integral constraint (5.24), this equation can be solved for the velocity fields $w_\pm$. The solution can be plugged into the stochastic differential equation

$$\begin{cases} d_\pm Z_\lambda^\mu & = w_\pm^\mu(Z_\lambda) d\lambda + \delta_\alpha^\mu dM_\lambda^\alpha, \\ d[M^\alpha, M^\beta]_\lambda & = \alpha \varepsilon \eta^{\alpha\beta} d\lambda, \end{cases}$$
(5.26)

which can be solved for the process $X = \text{Re}[Z]$.

We note that the velocity fields $w_\pm$ are also subjected to a relativistic generalization of the constraint (4.31), which is given by

$$\sin\left(\frac{\phi}{2}\right) u_+^0 + \cos\left(\frac{\phi}{2}\right) v_+^0 = \sin\left(\frac{\phi}{2}\right) u_-^0 + \cos\left(\frac{\phi}{2}\right) v_-^0,$$

$$\cos\left(\frac{\phi}{2}\right) u_+^i - \sin\left(\frac{\phi}{2}\right) v_+^i = \cos\left(\frac{\phi}{2}\right) u_-^i - \sin\left(\frac{\phi}{2}\right) v_-^i.$$
(5.27)

5.2 Diffusion Equations

The stochastic relativistic Hamilton-Jacobi equations (5.21) can be combined. Using the expression (5.22) for w_2, this yields the complex partial differential equation

$$-\frac{2}{\varepsilon}\frac{\partial S^{\pm}}{\partial \lambda} = \partial_{\mu}S^{\pm}\,\partial^{\mu}S^{\pm} \pm \alpha\,\partial_{\mu}\partial^{\mu}S^{\pm} - 2q\,A^{\mu}\,\partial_{\mu}S^{\pm} \mp \alpha q\,\partial_{\mu}A^{\mu} + q^2\,A_{\mu}A^{\mu} + m^2\,.$$

$$(5.28)$$

If we define the wave functions

$$\Psi_{\pm}(x,\varepsilon,\lambda) := \Phi_{\pm}(x)\exp\left(\pm\frac{\varepsilon\,m^2}{2\,\alpha}\lambda\right),$$

$$\Phi_{\pm}(x) := \exp\left(\pm\frac{S^{\pm}(x)}{\alpha}\right),$$

$$(5.29)$$

we find that $\Psi_{\pm}$ satisfy the complex diffusion equations

$$-\alpha\frac{\partial}{\partial\lambda}\Psi_{+} = \frac{\varepsilon}{2}\,\eta^{\mu\nu}\left(\alpha\frac{\partial}{\partial x^{\mu}} - q\,A_{\mu}\right)\left(\alpha\frac{\partial}{\partial x^{\nu}} - q\,A_{\nu}\right)\Psi_{+}\,,\qquad (5.30)$$

$$\alpha\frac{\partial}{\partial\lambda}\Psi_{-} = \frac{\varepsilon}{2}\,\eta^{\mu\nu}\left(\alpha\frac{\partial}{\partial x^{\mu}} + q\,A_{\mu}\right)\left(\alpha\frac{\partial}{\partial x^{\nu}} + q\,A_{\nu}\right)\Psi_{-}\,,\qquad (5.31)$$

while $\Phi_{\pm}$ satisfy the complex wave equations

$$\left[\eta^{\mu\nu}\left(\alpha\frac{\partial}{\partial x^{\mu}} - q\,A_{\mu}\right)\left(\alpha\frac{\partial}{\partial x^{\nu}} - q\,A_{\nu}\right) + m^2\right]\Phi_{+} = 0\,,\qquad (5.32)$$

$$\left[\eta^{\mu\nu}\left(\alpha\frac{\partial}{\partial x^{\mu}} + q\,A_{\mu}\right)\left(\alpha\frac{\partial}{\partial x^{\nu}} + q\,A_{\nu}\right) + m^2\right]\Phi_{-} = 0\,.\qquad (5.33)$$

For $\alpha = \pm i$, these wave equations reduce to the Klein-Gordon equation. As in previous chapters, there is a correspondence between the solutions of these equations and semi-martingale processes Z that solve the Itô equation (5.26) with velocity

$$w^{\mu}_{\pm} = \varepsilon\,\eta^{\mu\nu}\left(\pm\alpha\,\partial_{\nu}\ln\Psi_{\pm} - q\,A_{\nu}\right).\qquad (5.34)$$

Chapter 6
Stochastic Dynamics
on Pseudo-Riemannian Manifolds

Abstract This chapter extends the complex stochastic theory from Chaps. 4 and 5 to stochastic theories on Riemannian and Lorentzian manifolds. In doing so, it discusses the framework of second order geometry, which generalizes differential geometry to a stochastic context.

In the previous chapters, we have discussed stochastic dynamics on the Euclidean space $\mathbb{R}^d$ and the Minkowski space $\mathbb{R}^{d,1}$. In this chapter, we will extend the stochastic theory to the case, where the underlying space is no longer flat, but given by a pseudo-Riemannian manifold $(\mathcal{M}, g)$. Here, we will focus on relativistic theories on n-dimensional Lorentzian manifolds $(\mathcal{M}, g)$ with $n = d + 1$, but all the results in this chapter can be generalized in a straightforward manner to non-relativistic theories on d-dimensional Riemannian manifolds $(\mathcal{M}, g)$.

The main difference between a relativistic theory on $\mathbb{R}^{d,1}$ and a relativistic theory on the n-dimensional Lorentzian manifold $(\mathcal{M}, g)$ with $n = d + 1$ is that the Minkowski metric $\eta_{\mu\nu}$ is promoted to a smooth symmetric bilinear map $g_{\mu\nu} : T\mathcal{M} \times T\mathcal{M} \to \mathbb{R}$.

As a consequence, the classical Euler-Lagrange equations are modified to incorporate the curvature of spacetime and are given by

$$\frac{dX_\lambda^\mu}{d\lambda} = V_\lambda^\mu, \tag{6.1}$$

$$g_{\mu\nu}(X_\lambda)\left(\frac{dV_\lambda^\nu}{d\lambda} + \Gamma_{\rho\sigma}^\nu(X_\lambda)\, V_\lambda^\rho V_\lambda^\sigma\right) = \varepsilon\, q\, F_{\mu\nu}(X_\lambda)\, V_\lambda^\nu, \tag{6.2}$$

$$g_{\mu\nu}(X_\lambda)\, V_\lambda^\mu V_\lambda^\nu = -\varepsilon^2\, m^2, \tag{6.3}$$

where Γ is the Christoffel symbol associated to the Levi-Civita connection and where we use natural units $\hbar = c = 1$.

In the stochastic theory, we promote the trajectories $\{X_\lambda : \lambda \in \mathcal{T}\}$ to $\mathcal{M}$-valued semi-martingale processes. These are processes that can, in any local coordinate chart, be decomposed as

$$X_\lambda^\mu = C_\lambda^\mu + e_\alpha^\mu(X_\lambda)\, \mathrm{Re}[M_\lambda^\alpha], \tag{6.4}$$

where X_λ and C_λ are $\mathcal{M}$-valued and M_λ is $\mathbb{C}^n$-valued. Moreover, the polyads e^μ_α are now defined by the relation

$$g_{\mu\nu} e^\mu_\alpha e^\nu_\beta = \eta_{\alpha\beta} . \tag{6.5}$$

We will again impose the structure relation (5.7), i.e.

$$d[M^\alpha, M^\beta]_\lambda = \alpha \, \varepsilon \, \eta^{\alpha\beta} \, d\lambda$$

with $\alpha \in \mathbb{C}$, which implies

$$d[Z^\mu, Z^\nu]_\lambda = \alpha \, \varepsilon \, g^{\mu\nu}(X_\lambda) \, d\lambda + o(d\lambda) \tag{6.6}$$

with $Z_\lambda = X_\lambda + i\, Y_\lambda$ and Y_λ an auxiliary $\mathcal{M}$-valued process.

6.1 Second Order Phase Space

The major difficulty in extending the stochastic theory to a manifold resides in the fact that the velocity fields

$$w^\mu_+(X_\lambda) = \lim_{d\lambda \to 0} \mathbb{E}\left[\frac{Z^\mu_{\lambda+d\lambda} - Z^\mu_\lambda}{d\lambda} \,\middle|\, X_\lambda \right] \tag{6.7}$$

$$w^\mu_-(X_\lambda) = \lim_{d\lambda \to 0} \mathbb{E}\left[\frac{Z^\mu_\lambda - Z^\mu_{\lambda-d\lambda}}{d\lambda} \,\middle|\, X_\lambda \right] \tag{6.8}$$

$$w^{\mu\nu}_2(X_\lambda) = \lim_{d\lambda \to 0} \mathbb{E}\left[\frac{[Z^\mu_{\lambda+d\lambda} - Z^\mu_\lambda][Z^\nu_{\lambda+d\lambda} - Z^\nu_\lambda]}{h} \,\middle|\, X_\lambda \right] \tag{6.9}$$

are well defined in every coordinate chart containing X_λ, but the fields $w_\pm$ are not covariant. This issue can be resolved using second order geometry, cf. Appendix D. In second order geometry, the n-dimensional tangent spaces $T_x\mathcal{M}$ at every point x are extended to $\frac{1}{2}n(n+3)$-dimensional tangent spaces $T_{\pm,x}\mathcal{M}$. Moreover, given a coordinate chart, any vector $w_\pm \in T_{\pm,x}\mathcal{M}$ can be represented with respect to the canonical basis of $T_{\pm,x}\mathcal{M}$ as

$$w_\pm = w^\mu_\pm \, \partial_\mu \pm \frac{1}{2} w^{\mu\nu}_2 \, \partial_{\mu\nu} . \tag{6.10}$$

This allows to construct covariant objects

$$\hat{w}^\mu_\pm = w^\mu_\pm \pm \frac{1}{2} \Gamma^\mu_{\nu\rho} w^{\nu\rho}_2 , \tag{6.11}$$

such that the fields $\hat{w}_{\pm}^{\mu}$ transform in a covariant manner. We note that $\hat{w}_{\mathrm{o}} = w_{\mathrm{o}}$, which reflects the fact that the Stratonovich field is a first order vector field and thus transforms covariantly. Therefore, at any point $x \in \mathcal{M}$, one can define a Stratonovich tangent space $T_{\mathrm{o},x}\mathcal{M}$, and, given a coordinate chart, vectors $w_{\mathrm{o}} \in T_{\mathrm{o},x}\mathcal{M}$ can be represented with respect to the canonical basis of $T_{\mathrm{o},x}\mathcal{M}$ as

$$w_{\mathrm{o}} = w_{\mathrm{o}}^{\mu}\, \partial_{\mu}\,. \tag{6.12}$$

The Stratonovich and Itô tangent bundles can be constructed from the respective tangent spaces, such that

$$T_{\mathrm{o}}\mathcal{M} = \bigsqcup_{x \in \mathcal{M}} T_{\mathrm{o},x}\mathcal{M}\,,$$

$$T_{\pm}\mathcal{M} = \bigsqcup_{x \in \mathcal{M}} T_{\pm,x}\mathcal{M}\,.$$

This allows to study processes $(X_{\lambda}, W_{\mathrm{o},\lambda})$ on $T_{\mathrm{o}}^{\mathbb{C}}\mathcal{M}$ and $(X_{\lambda}, W_{\pm,\lambda}, \pm W_{2,\lambda})$ on $T_{\pm}^{\mathbb{C}}\mathcal{M}$ that satisfy[1]

$$W_{\mathrm{o},\lambda}^{\mu}\, d\lambda = d_{\mathrm{o}} Z_{\lambda}^{\mu}\,,$$

$$W_{\pm,\lambda}^{\mu}\, d\lambda = d_{\pm} Z_{\lambda}^{\mu}\,,$$

$$W_{2,\lambda}^{\mu\nu}\, d\lambda = d[Z^{\mu}, Z^{\nu}]_{\lambda}\,.$$

Alternatively, one can express the processes on $T_{\pm}^{\mathbb{C}}\mathcal{M}$ in an explicitly covariant form $(X_{\lambda}, \hat{W}_{\pm,\lambda}, \pm W_{2,\lambda})$, where

$$\hat{W}_{\pm,\lambda}^{\mu} = W_{\pm,\lambda}^{\mu} \pm \frac{1}{2}\Gamma_{\nu\rho}^{\mu}(X_{\lambda})\, W_{2,\lambda}^{\nu\rho} \tag{6.13}$$

and

$$\hat{W}_{\pm,\lambda}^{\mu}\, d\lambda = d_{\pm} Z_{\lambda}^{\mu} \pm \frac{1}{2}\Gamma_{\nu\rho}^{\mu}(X_{\lambda})\, d[Z^{\nu}, Z^{\rho}]_{\lambda}\,.$$

6.2   Equations of Motion

Using second order geometry, all results from Chaps. 3, 4 and 5 can be generalized to pseudo-Riemannian manifolds. In this section, we state the results for the generalization of the complex relativistic theory discussed in Chap. 5.

---

[1] These differential expressions are defined by an integral expression as in Eq. (3.27).

The Stratonovich Lagrangian is given by

$$L^\circ(x, w_\circ, \varepsilon) = \frac{1}{2\,\varepsilon}\, g_{\mu\nu}(x)\, w_\circ^\mu w_\circ^\nu - \frac{\varepsilon\, m^2}{2} + q\, A_\mu(x)\, w_\circ^\mu \tag{6.14}$$

and obeys the Stratonovich-Euler-Lagrange equations

$$d_\circ \frac{\partial L^\circ}{\partial w_\circ^\mu} = \frac{\partial L^\circ}{\partial x^\mu}\, d\lambda, \tag{6.15}$$

$$\mathbb{E}\left[\frac{\partial L^\circ}{\partial \varepsilon}\right] = 0. \tag{6.16}$$

For the Lagrangian (6.14), this yields

$$d_\circ Z^\mu = W_\circ^\mu\, d\lambda, \tag{6.17}$$

$$g_{\mu\nu}\left(d_\circ W_\circ^\nu + \Gamma^\nu_{\rho\sigma}\, W_\circ^\rho W_\circ^\sigma\, d\lambda\right) = \varepsilon\, q\, F_{\mu\nu}\, W_\circ^\nu\, d\lambda, \tag{6.18}$$

$$\mathbb{E}\left[g_{\mu\nu}\, W_\circ^\mu W_\circ^\nu\right] = -\varepsilon^2\, m^2, \tag{6.19}$$

where the last equation is the energy-momentum relation, which can be rewritten as the stochastic line element

$$\mathbb{E}\left[g_{\mu\nu}\, d_\circ X^\mu\, d_\circ X^\nu\right] = -\varepsilon^2\, m^2\, d\lambda^2. \tag{6.20}$$

The generalization of the Itô formulation to manifolds is less straightforward. The Itô Lagrangian is derived in Appendix E, where it is found that

$$L^\pm(x, w_\pm, w_2, \varepsilon) = L_0^\pm(x, w_\pm, w_2, \varepsilon) \pm L_\infty(x, w_\circ, \varepsilon) \tag{6.21}$$

with finite part

$$\begin{aligned}
L_0^\pm(x, w_\pm, w_2, \varepsilon) = {} & \frac{1}{2\,\varepsilon}\, g_{\mu\nu}(x)\left(w_\pm^\mu \pm \frac{1}{2}\Gamma^\mu_{\rho\sigma}(x)\, w_2^{\rho\sigma}\right)\left(w_\pm^\nu \pm \frac{1}{2}\Gamma^\nu_{\kappa\lambda}(x)\, w_2^{\kappa\lambda}\right) \\
& + \frac{1}{12\,\varepsilon}\, \mathcal{R}_{\mu\nu\rho\sigma}(x)\, w_2^{\mu\rho} w_2^{\nu\sigma} - \frac{\varepsilon\, m^2}{2} \\
& + q\, A_\mu\left(w_\pm^\mu \pm \frac{1}{2}\Gamma^\mu_{\nu\rho}(x)\, w_2^{\nu\rho}\right) \pm \frac{q}{2}\, \nabla_\nu A_\mu\, w_2^{\mu\nu} \tag{6.22}
\end{aligned}$$

and a divergent part such that

$$\mathbb{E}\left[\int L_\infty^\pm(x, w_\circ, \varepsilon)\, d\lambda\right] = \pm\, \mathbb{E}\left[\int \frac{1}{2\,\varepsilon}\, g_{\mu\nu}(x)\, d[z^\mu, w_\circ^\nu]\right]. \tag{6.23}$$

The corresponding Itô-Euler-Lagrange equations are given by, cf. Appendix F.2,

$$d_\pm \frac{\partial L_0^\pm}{\partial w_\pm^\mu} = \frac{\partial L_0^\pm}{\partial x^\mu} d\lambda - \Gamma_{\mu\nu}^\rho \left(\frac{\partial L_0^\pm}{\partial w_2^{\rho\sigma}} + \frac{\partial L_0^\pm}{\partial w_2^{\sigma\rho}} \right) d[Z^\nu, Z^\sigma] \pm \Gamma_{\mu\nu}^\rho \frac{\partial^2 L_0^\pm}{\partial x^\sigma \partial w_\pm^\rho} d[Z^\nu, Z^\sigma]$$

$$\pm \Gamma_{\mu\nu}^\rho \frac{\partial^2 L_0^\pm}{\partial w_\pm^\sigma \partial w_\pm^\rho} d[Z^\nu, W_\pm^\sigma] \pm \Gamma_{\mu\nu}^\rho \frac{\partial^2 L_0^\pm}{\partial w_2^{\sigma\kappa} \partial w_\pm^\rho} d[Z^\nu, W_2^{\sigma\kappa}], \tag{6.24}$$

$$\mathbb{E}\left[\frac{\partial L_0^\pm}{\partial \varepsilon} \right] = 0. \tag{6.25}$$

For the Lagrangian (6.22), this yields

$$d_\pm Z^\mu \pm \Gamma_{\nu\rho}^\mu d[Z^\nu, Z^\rho] = \hat{W}_\pm^\mu d\lambda, \tag{6.26}$$

$$d[Z^\mu, Z^\nu] = W_2^{\mu\nu} d\lambda, \tag{6.27}$$

$$\text{LHS} = \text{RHS}, \tag{6.28}$$

$$\mathbb{E}\left[g_{\mu\nu} \hat{W}_\pm^\mu \hat{W}_\pm^\nu + \frac{1}{6} \mathcal{R}_{\mu\nu\rho\sigma} W_2^{\mu\rho} W_2^{\nu\sigma} \right] = -\varepsilon^2 m^2 \tag{6.29}$$

with

$$\text{LHS} = g_{\mu\nu} d_\pm \hat{W}_\pm^\nu + g_{\mu\nu} \Gamma_{\rho\sigma}^\nu \hat{W}_\pm^\rho \left(\hat{W}_\pm^\sigma - \frac{1}{2} \Gamma_{\kappa\lambda}^\sigma W_2^{\kappa\lambda} \right) d\lambda \pm g_{\mu\nu} \Gamma_{\rho\sigma}^\nu d[Z^\rho, \hat{W}_\pm^\sigma]$$

$$\pm \frac{1}{2} \left[\mathcal{R}_{\mu\rho\sigma\nu} + g_{\mu\kappa} \left(\partial_\rho \Gamma_{\sigma\nu}^\kappa + \Gamma_{\rho\lambda}^\kappa \Gamma_{\sigma\nu}^\lambda \right) \right] \hat{W}_2^\nu W_2^{\rho\sigma} d\lambda, \tag{6.30}$$

$$\text{RHS} = \varepsilon q F_{\mu\nu} \hat{W}_\pm^\nu d\lambda \pm \frac{\varepsilon q}{2} \nabla_\rho F_{\mu\nu} W_2^{\nu\rho} d\lambda + \frac{1}{12} \nabla_\mu \mathcal{R}_{\nu\rho\sigma\kappa} W_2^{\nu\sigma} W_2^{\rho\kappa} d\lambda. \tag{6.31}$$

Moreover, the energy-momentum relation (6.29) can be rewritten as the stochastic line element

$$\mathbb{E}\left[g_{\mu\nu} \left(d_\pm \hat{X}^\mu d_\pm \hat{X}^\nu \mp d[X^\mu, X^\nu] \right) + \frac{1}{6} \mathcal{R}_{\mu\nu\rho\sigma} d[X^\mu, X^\rho] d[X^\nu, X^\sigma] \right] = -\varepsilon^2 m^2 d\lambda^2. \tag{6.32}$$

6.3 Hamilton-Jacobi Equations

Hamilton's principal functions are complex valued functions given by

$$S^+(x, \varepsilon, \lambda) = -\mathbb{E}\left[\int_\lambda^{\lambda_f} L^+(X_\tau, W_{+,\tau}, W_{2,\tau}, \mathcal{E}_\tau) d\tau \, \middle| \, X_\lambda = x, X_{\lambda_f} = x_f, \mathcal{E}_\lambda = \varepsilon, \mathcal{E}_{\lambda_f} = \varepsilon_f \right]$$

$$S^-(x, \varepsilon, \lambda) = \mathbb{E}\left[\int_{\lambda_0}^\lambda L^-(X_\tau, W_{-,\tau}, W_{2,\tau}, \mathcal{E}_\tau) d\tau \, \middle| \, X_\lambda = x, X_{\lambda_0} = x_0, \mathcal{E}_\lambda = \varepsilon, \mathcal{E}_{\lambda_0} = \varepsilon_0 \right]$$

$$\tag{6.33}$$

and can be used to derive the Hamilton-Jacobi equations. These are derived in Appendix F.3 and given by

$$
\begin{cases}
\frac{\partial}{\partial x^\mu} S^\pm(x, \varepsilon, \lambda) & = p_\mu^\pm, \\
\left(\frac{\partial}{\partial \lambda} + \frac{d\varepsilon(\lambda)}{d\lambda} \frac{\partial}{\partial \varepsilon} \right) S^\pm(x, \varepsilon, \lambda) & = -p_\mu^\pm \hat{w}_\pm^\mu \mp \frac{1}{2} \nabla_\nu p_\mu^\pm \, w_2^{\mu\nu} + L_0^\pm(x, w_\pm, w_2, \varepsilon),
\end{cases}
$$
$$(6.34)$$

where the momentum $p^\pm$ is complex valued, and w_2 is determined by Eq. (6.6), i.e.

$$
w_2^{\mu\nu}(X_\lambda) = \alpha \, \varepsilon \, g^{\mu\nu}(X_\lambda).
$$
$$(6.35)$$

Furthermore, the reparameterization invariance of the theory, given in Eq. (5.2), implies the constraint

$$
\frac{\partial S^\pm}{\partial \lambda} + \frac{d\varepsilon}{d\lambda} \frac{\partial S^\pm}{\partial \varepsilon} = 0.
$$
$$(6.36)$$

In addition, the velocity field $\hat{w}_\pm(X_\lambda)$ satisfies an integral constraint given by

$$
\oint \left(p_\mu^\pm \, \hat{w}_\pm^\mu \pm \frac{1}{2} w_2^{\mu\nu} \, \nabla_\nu p_\mu^\pm \right) d\lambda = \pm \alpha \, \pi \, \mathrm{i} k_\mu^\mu,
$$
$$(6.37)$$

where $k_\nu^\mu \in \mathbb{Z}^{n \times n}$ is a matrix of winding numbers.

If we take a covariant derivative of the second Hamilton-Jacobi equation, we find

$$
g_{\nu\rho} \hat{w}_\pm^\rho \nabla_\mu \hat{w}_\pm^\nu \pm \frac{1}{2} g_{\nu\rho} w_2^{\rho\sigma} \nabla_\mu \nabla_\sigma \hat{w}_\pm^\nu = \frac{1}{12} w_2^{\nu\sigma} w_2^{\rho\kappa} \nabla_\mu \mathcal{R}_{\nu\rho\sigma\kappa}.
$$
$$(6.38)$$

Then, by applying Eq. (2.18) in the form

$$
g_{\nu\rho} \nabla_\mu \hat{w}_\pm^\nu = g_{\mu\nu} \nabla_\rho \hat{w}_\pm^\nu - \varepsilon q \, F_{\mu\rho},
$$
$$(6.39)$$

we obtain

$$
\left[g_{\mu\nu} \hat{w}_\pm^\rho \nabla_\rho \pm \frac{1}{2} w_2^{\rho\sigma} \left(g_{\mu\nu} \nabla_\rho \nabla_\sigma + \mathcal{R}_{\mu\rho\sigma\nu} \right) - \varepsilon q \, F_{\mu\nu} \right] \hat{w}_\pm^\nu = \frac{1}{12} w_2^{\nu\sigma} w_2^{\rho\kappa} \nabla_\mu \mathcal{R}_{\nu\rho\sigma\kappa}
$$
$$
\pm \frac{\varepsilon q}{2} w_2^{\nu\rho} \nabla_\rho F_{\mu\nu},
$$
$$(6.40)$$

which is equivalent to Eq. (6.28) with W_λ replaced by $w(X_\lambda)$.

Taking into account the integral constraint (6.37), Eq. (6.40) can be solved for the velocity fields $\hat{w}_\pm$. The solution can be plugged into the stochastic differential equation

$$
\begin{cases}
d_\pm Z_\lambda^\mu & = \left(\hat{w}_\pm^\mu \mp \frac{1}{2} \Gamma_{\nu\rho}^\mu w_2^{\nu\rho} \right) d\lambda + e_\alpha^\mu dM_\lambda^\alpha, \\
d[M^\alpha, M^\beta]_\lambda & = \alpha \varepsilon \delta^{\alpha\beta} d\lambda,
\end{cases}
$$
$$(6.41)$$

which can be solved for Z. The fields $w_\pm$ are subjected to the constraint (5.27), which reflects that the boundary conditions for w_+ and w_- cannot be chosen completely independently.

6.4   Diffusion Equations

Combining the stochastic Hamilton-Jacobi equations (6.34) and plugging in expression (6.35) for w_2 yields a complex partial differential equation

$$-\frac{2}{\varepsilon}\frac{\partial S^\pm}{\partial\lambda} = \nabla_\mu S^\pm \nabla^\mu S^\pm \pm \alpha\Box S^\pm - 2q\,A^\mu\,\nabla_\mu S^\pm \mp \alpha q\,\nabla_\mu A^\mu + q^2\,A_\mu A^\mu - \frac{\alpha^2}{6}\mathcal{R} + m^2.$$
(6.42)

If we define the wave functions (5.29):

$$\Psi_\pm(x,\varepsilon,\lambda) := \Phi(x)\exp\left(\pm\frac{\varepsilon\,m^2}{2\,\alpha}\lambda\right),$$

$$\Phi_\pm(x) := \exp\left(\pm\frac{S^\pm(x)}{\alpha}\right),$$

we find that $\Psi_\pm$ satisfy the complex diffusion equations

$$-\alpha\frac{\partial}{\partial\lambda}\Psi_+ = \frac{\varepsilon}{2}\left[\left(\alpha\nabla_\mu - q\,A_\mu\right)\left(\alpha\nabla^\mu - q\,A^\mu\right) - \frac{\alpha^2}{6}\mathcal{R}\right]\Psi_+,$$
(6.43)

$$\alpha\frac{\partial}{\partial\lambda}\Psi_- = \frac{\varepsilon}{2}\left[\left(\alpha\nabla_\mu + q\,A_\mu\right)\left(\alpha\nabla^\mu + q\,A^\mu\right) - \frac{\alpha^2}{6}\mathcal{R}\right]\Psi_-,$$
(6.44)

while $\Phi_\pm$ satisfy the complex wave equations

$$\left[\left(\alpha\nabla_\mu - q\,A_\mu\right)\left(\alpha\nabla^\mu - q\,A^\mu\right) - \frac{\alpha^2}{6}\mathcal{R} + m^2\right]\Phi_+ = 0,$$
(6.45)

$$\left[\left(\alpha\nabla_\mu + q\,A_\mu\right)\left(\alpha\nabla^\mu + q\,A^\mu\right) - \frac{\alpha^2}{6}\mathcal{R} + m^2\right]\Phi_- = 0.$$
(6.46)

For $\alpha = \pm i$ these equations reduce to the Klein-Gordon equation including the Pauli-DeWitt term [108, 109], which represents a non-minimal coupling $\xi = \frac{1}{6}$ to the Ricci scalar. For $n = 4$, this coupling coincides with the conformal coupling $\xi = \frac{n-2}{4(n-1)}$, but this is not true for any other dimension.

6.5 Spacetime Symmetries

Any non-relativistic theory is locally invariant under the action of the Galilean group $\text{Gal}(d)$, which can be written as a Cartesian product

$$\text{Gal}(d) = O(d) \times \mathbb{R}^d \times \mathbb{R}^d \times \mathbb{R} \tag{6.47}$$

with group multiplication

$$(R', v', a', s')\,(R, v, a, s) = (R'\,R,\ R'\,v + v',\ R'\,a + a' + v'\,s,\ s' + s) \tag{6.48}$$

for all $R, R' \in O(d)$, $v, v' \in (\mathbb{R}^d, +)$, $a, a' \in (\mathbb{R}^d, +)$ and $s, s' \in (\mathbb{R}, +)$. The left action of this group on the space $\mathbb{R}^{d+1}$ is given by

$$(R, v, a, s)\,(x, t) = (R\,x + v\,t + a,\ t + s) \tag{6.49}$$

for all $(R, v, a, s) \in \text{Gal}(d)$ and $(x, t) \in \mathbb{R}^{d+1}$.

Relativistic theories, on the other hand, are locally invariant under the action of the Poincaré group $\text{IO}(d, 1)$, which can be written as a semi-direct product

$$\text{IO}(d, 1) = O(d, 1) \ltimes \mathbb{R}^{d,1}. \tag{6.50}$$

with group multiplication

$$(\Lambda', b')\,(\Lambda, b) = (\Lambda'\,\Lambda,\ \Lambda'\,b + b') \tag{6.51}$$

for all $\Lambda, \Lambda' \in O(d, 1)$ and $b, b' \in (\mathbb{R}^{d,1}, +)$. The left action of this group on $\mathbb{R}^{d,1}$ is given by

$$(\Lambda, b)\,x = \Lambda\,x + b \tag{6.52}$$

for all $(\Lambda, b) \in \text{IO}(d, 1)$ and $x \in \mathbb{R}^{d,1}$.

In stochastic theories on curved spacetime the Galilean symmetry and Poincaré symmetry are no longer the correct symmetries of spacetime, since the structure group of the (co)tangent bundle is deformed, cf. Appendix D.2.

In a non-relativistic theory, the structure group of the tangent bundle is $GL(d, \mathbb{R})$, which has the orthogonal group $O(d)$ as a subgroup. In a stochastic theory, the structure group of the Itô tangent bundle is the Itô group G_I^d, which can be written as a Cartesian product

$$G_I^d = GL(d, \mathbb{R}) \times \text{Lin}(\mathbb{R}^d \otimes \mathbb{R}^d, \mathbb{R}^d) \tag{6.53}$$

with group multiplication

$$(g', \kappa')\,(g, \kappa) = (g'\,g,\ g' \circ \kappa + \kappa' \circ (g \otimes g)) \tag{6.54}$$

for all $g, g' \in GL(d, \mathbb{R})$ and $\kappa, \kappa' \in Lin(\mathbb{R}^d \otimes \mathbb{R}^d, \mathbb{R}^d)$. Furthermore, its left action on $\mathbb{R}^d \times Sym(T\mathbb{R}^d \otimes T\mathbb{R}^d)$ is given by

$$(g, \kappa)(x, x_2) = (g\,x + \kappa\,x_2, (g \otimes g)\,x_2) \tag{6.55}$$

for all $(g, \kappa) \in G_I^d$, $x \in \mathbb{R}^d$ and $x_2 \in Sym(T\mathbb{R}^d \otimes T\mathbb{R}^d)$.

This Itô deformation of the $GL(d, \mathbb{R})$ is inherited by the orthogonal group, which induces a deformation of the Galilean group to $Gal_I(d)$. We can define this group as a Cartesian product

$$Gal_I(d) = Gal(d) \times Lin(\mathbb{R}^d \otimes \mathbb{R}^d, \mathbb{R}^d) \tag{6.56}$$

with group multiplication

$$\begin{aligned}
(R', v', a', s', \kappa')\,(R, v, a, s, \kappa) \\
= (R'\,R,\ R'v + v',\ R'a + a' + v's,\ s' + s,\ R' \circ \kappa + \kappa' \circ (R \otimes R))
\end{aligned} \tag{6.57}$$

for all $R, R' \in O(d)$, $v, v' \in (\mathbb{R}^d, +)$, $a, a' \in (\mathbb{R}^d, +)$, $s, s' \in (\mathbb{R}, +)$ and $\kappa, \kappa' \in Lin(\mathbb{R}^d \otimes \mathbb{R}^d, \mathbb{R}^d)$. The left action of this group on the space $\mathbb{R}^{d+1} \times Sym(T\mathbb{R}^d \otimes T\mathbb{R}^d)$ is given by

$$(R, v, a, s, \kappa)(x, t, x_2) = (R\,x + v\,t + a + \kappa\,x_2,\ t + s,\ (R \otimes R)\,x_2) \tag{6.58}$$

for all $(R, v, a, s, \kappa) \in Gal_I(d)$, $(x, t) \in \mathbb{R}^{d+1}$ and $x_2 \in Sym(T\mathbb{R}^d \otimes T\mathbb{R}^d)$.

A similar analysis can be applied to relativistic theories. For this, we first construct a relativistic version of the Itô group given by the Cartesian product

$$G_I^{d,1} = GL(d + 1, \mathbb{R}) \times Lin(\mathbb{R}^{d,1} \otimes \mathbb{R}^{d,1}, \mathbb{R}^{d,1}) \tag{6.59}$$

with group multiplication as given in Eq. (6.54) and a left action on $\mathbb{R}^{d,1} \times Sym(T\mathbb{R}^{d,1} \otimes T\mathbb{R}^{d,1})$ as given in Eq. (6.55). This Itô deformation of $GL(d + 1, \mathbb{R})$ is inherited by the Lorentz group $O(d, 1)$, which induces a deformation of the Poincaré group to $IO_I(d, 1)$. We define this group as the Cartesian product

$$IO_I(d, 1) = IO(d, 1) \times Lin(\mathbb{R}^{d,1} \otimes \mathbb{R}^{d,1}, \mathbb{R}^{d,1}) \tag{6.60}$$

with group multiplication

$$(\Lambda', b', \kappa')\,(\Lambda, b, \kappa) = (\Lambda'\,\Lambda,\ \Lambda'b + b',\ \Lambda' \circ \kappa + \kappa' \circ (\Lambda \otimes \Lambda)) \tag{6.61}$$

for all $\Lambda, \Lambda' \in O(d, 1)$, $b, b' \in (\mathbb{R}^{d,1}, +)$ and $\kappa, \kappa' \in Lin(\mathbb{R}^{d,1} \otimes \mathbb{R}^{d,1}, \mathbb{R}^{d,1})$. The left action of this group on the space $\mathbb{R}^{d,1} \times Sym(T\mathbb{R}^{d,1} \otimes T\mathbb{R}^{d,1})$ is then given by

$$(\Lambda, b, \kappa)(x, x_2) = (\Lambda\,x + b + \kappa\,x_2, (\Lambda \otimes \Lambda)\,x_2) \tag{6.62}$$

for all $(\Lambda, b, \kappa) \in IO_I(d, 1)$, $(x) \in \mathbb{R}^{d,1}$ and $x_2 \in \mathrm{Sym}(T\mathbb{R}^{d,1} \otimes T\mathbb{R}^{d,1})$.

We conclude this section with two remarks. First, the Itô deformations of the space-time symmetries only apply to the Itô formulation of the theory. In the Stratonovich framework the symmetries remain undeformed.

Secondly, the Itô deformations of the spacetime symmetries vanish in the limits $\kappa \to 0$ and $x_2 \to 0$. In this book, these limits have a clear physical interpretation, as we have fixed the quadratic variation to be proportional to the inverse metric. Due to this choice, we have

- $x_2 = \frac{\alpha \hbar}{m} g t$ and $\kappa = \Gamma$ for non-relativistic theories,
- $x_2 = \alpha \hbar \varepsilon g \lambda$ and $\kappa = \Gamma$ for relativistic theories,

where Γ is the Levi-Civita connection associated to the metric g. Consequently, the Itô deformations vanish in the classical limit $\hbar \to 0$, as this implies $x_2 \to 0$, and in the limit of vanishing gravity $G \to 0$, as this implies $\Gamma \to 0$.

Chapter 7
Stochastic Interpretation

Abstract This chapter discusses various interpretational aspects of the stochastic theory in more detail. It focuses on the notions of locality and causality in the theory, and it provides an elementary analysis of the Bell inequalities for stochastic theories. In addition, it provides a possible quantum gravity inspired interpretation for the source of the stochastic fluctuations.

The theory that we have presented in this book is inherently stochastic and does not contain any deterministic hidden variables. Instead, the theory should be regarded as a mathematically rigorous implementation of the statement 'God plays dice'. Here, the dice game is played in the probability space (Ω, Σ), which determines the possible outcomes of the dice rolls, and the probability for each of these outcomes is determined by the probability measure $\mathbb{P}$.

This stochastic theory describes a generalization of both quantum mechanics, which is obtained for $\phi = \frac{\pi}{2}$ and the theory of Brownian motion, which is obtained for $\phi = 0$. In physics, this mathematical theory of Brownian motion is only used as an effective theory to replace the more fundamental physical theory of Brownian motion. In this fundamental theory, the Brownian motion is not induced by a dice roll, but by the interaction of the macroscopic particle under consideration with a large number of microscopic particles. These interactions are fully deterministic, but their average effect can be modeled more efficiently by the stochastic theory. Therefore, in this physical picture, the probability measure $\mathbb{P}$ can in principle be derived from the underlying deterministic physics.

The above discussion suggests that stochastic mechanics may be reinterpreted as a hidden variable theory. In such a hidden variable interpretation, the stochastic law models the effective interactions of the particle with an omnipresent background field. One might wonder whether the dynamics of this background can itself be described by some (non-)deterministic physical laws. In this chapter, we explore this possibility by discussing various aspects of the stochastic theory in more detail. Moreover, we show that the Bell inequalities do not provide any objection to such a hidden variable theory.

F. Kuipers, *Stochastic Mechanics*, SpringerBriefs in Physics,
https://doi.org/10.1007/978-3-031-31448-3_7

7.1 Locality

The stochastic theory is a local theory, since it is governed by local physical laws of motion. The drift of the stochastic theory obeys a stochastic generalization of the principle of least action, which is a local principle. Moreover, the stochastic fluctuations of the theory are introduced by a local structure relation. The locality of the theory is reflected by the fact that the equations of motion that govern the motion of a stochastic particle are stochastic differential equations, and these are intrinsically local.

We emphasize that this notion of locality only applies to the stochastic theory of a single particle discussed in this book. It is expected that, in order to explain entanglement in multi-particle systems, the notion of locality in the stochastic theory must be relaxed. For example, for two particles, described by semi-martingales $X_{1,2}^{\mu} = C_{1,2}^{\mu} + e_{\alpha}^{\mu} M_{1,2}^{\alpha}$, the entanglement will introduce a stochastic coupling such that $d[M_1, M_2] \neq 0$. If this coupling is preserved, when X_1 and X_2 are separated, this coupling will introduce a non-locality in the stochastic law that governs this two particle state.

Furthermore, we note that, even though the single particle theory is local, we have defined objects that are not local. These are the wave functions $\Psi_{\pm}$ and the velocity fields $w_{\pm}$. The wave functions are solutions of the diffusion equations and therefore depend on the global properties of the manifold and the scalar and vector potential. A similar reasoning applies to the velocity fields, which are solutions of another set of partial differential equations and closely related to the wave functions.

These non-local objects do not render the theory non-local, as they do not represent physical properties of the particle. Only $w_{\pm}(X_t, t)$ along the trajectory X_t has a physical meaning, as it represents the expected velocity of the particle along its trajectory. The extension of $w(X_t, t)$ to $w(x, t)$ for all $x \in \mathcal{M}$ is introduced for mathematical convenience, but does not represent physical properties.

Since $w_{\pm}(x, t)$ is related to $\Psi_{\pm}(x, t)$ the same reasoning can be applied to the wave functions. This implies that the extension of the wave function $\Psi(x, t)$ to $x \in \mathcal{M}$ does not represent any intrinsic properties of the particle. Instead, it provides the best possible prediction that an observer can make about the particle. In making this prediction, observers use their knowledge about an initial condition $\Psi_-(X_0, t_0)$ or terminal condition $\Psi_+(X_f, t_f)$ and about the global properties of the manifold and the scalar and vector potential.

We conclude that the wave function is not intrinsic to the particle, but only describes how an observer sees the particle.[1] Therefore, the wave function is not a universal object. Due to this fact, wave function collapse does not introduce any peculiarities in the stochastic formulation. In the stochastic theory, a wave function collapse occurs when an observer updates its own knowledge about the particle after making a measurement. This leads to a collapse in the wave function that the observer uses to describe the particle, but does not affect the particle under consideration.

---

[1] This interpretation is standard in the study of Brownian motion. The complex stochastic theory shows that this interpretation of the wave function must also be adopted in quantum mechanics.

Another consequence of this non-universal character is that the wave function is an observer dependent object: different observers can describe the same particle using a different wave function, as their knowledge about the state of the particle might differ.

7.2 Causality

The stochastic theory is by construction causal with respect to any affine parameter, since both the deterministic and stochastic laws are defined with respect to an affine parameter. In a non-relativistic theory, the notion of time is an affine parameter. Therefore, the non-relativistic stochastic theory is causal with respect to time.

In a relativistic theory, time is no longer an affine parameter, as it is promoted to a coordinate in the spacetime. Therefore, one must impose further conditions to obtain causality with respect to time. In relativistic theories, causality with respect to time is obtained by imposing that all trajectories are causal. In a stochastic theory, this can be defined as follows[2]:

- a trajectory X_λ is causal, if $\mathbb{E}\left[g(v_\circ, v_\circ)\right] \leq 0$ along X_λ for all $\lambda \in \mathcal{T}$.

In practice, causality is ensured by imposing that all particles are time-like or null-like. In a stochastic theory, these notions can be defined by[3]

- a particle is time-like, if $\mathbb{E}\left[g(v_\circ, v_\circ)\right] < 0$ along its trajectory X_λ for all $\lambda \in \mathcal{T}$;
- a particle is null-like, if $\mathbb{E}\left[g(v_\circ, v_\circ)\right] = 0$ along its trajectory X_λ for all $\lambda \in \mathcal{T}$;
- a particle is space-like, if $\mathbb{E}\left[g(v_\circ, v_\circ)\right] > 0$ along its trajectory X_λ for all $\lambda \in \mathcal{T}$.

As in the deterministic theory, we can ensure that all particles fall into one of these categories by imposing an energy momentum relation or on-shell condition. In the stochastic theory, this condition is given by

$$
\mathbb{E}\left[g(v_\circ, v_\circ)\right] = -\varepsilon^2 m^2 = \begin{cases} -1 & \text{if } m^2 > 0, \\ 0 & \text{if } m^2 = 0, \\ +1 & \text{if } m^2 < 0, \end{cases} \tag{7.1}
$$

where we gauge fixed ε in the second equality. This condition can also be written in the Itô formulation using that

$$
\mathbb{E}\left[g(v_\circ, v_\circ)\right] = \mathbb{E}\left[g(\hat{v}_\pm, \hat{v}_\pm) + \frac{1}{6}\mathcal{R}_{\mu\nu\rho\sigma}v_2^{\mu\rho}v_2^{\nu\sigma}\right], \tag{7.2}
$$

which follows from Sect. 6.2 and Appendix E. If an on-shell condition is imposed, one finds that

---

[2] Note that this definition depends on the choice for the $(-+\cdots+)$ signature.
[3] Cf. Appendix B.9.

- a particle is time-like, if $m^2 > 0$;
- a particle is null-like, if $m^2 = 0$;
- a particle is space-like, if $m^2 < 0$.

Therefore, the theory is causal, if the rest mass of all particles is fixed to be $m^2 \geq 0$.

We point out that the stochastic notion of causality is slightly weaker than the classical notion of causality. In a classical theory, causality is defined without taking an expectation value, while in a stochastic theory the expectation value is necessary to make the causality condition well defined. As a consequence, stochastic causality ensures that the expectation value $\mathbb{E}[X] = \int_\Omega X(\omega)\, d\mathbb{P}(\omega)$ satisfies the classical notion of causality, but the sample paths $X(\omega)$ can violate this stricter classical notion of causality, as they can go off-shell.

For massless particles, deviations from the on-shell condition $g(v_\circ, v_\circ) = 0$, immediately imply that the particle is no longer null-like, and, therefore, that the classical causality condition is violated. For massive particles, on the other hand, deviations from the on-shell condition $g(v, v) = -1$, do not imply violation of the classical causality condition, as they may retain their time-like character. Nevertheless, for large deviations from the on-shell condition a violation of classical causality may occur.

We can provide an estimate for the likelihood of such violations by considering a free massive particle in a flat spacetime. Its trajectory is a solution of the Itô equation

$$\begin{cases} d_+ X_\tau^\mu & = v_+^\mu(X_\tau, \tau)\, d\tau + \mathrm{Re}[dM_\tau^\mu]\,, \\ d[M^\mu, M^\nu]_\tau & = \frac{\hbar}{m}\, \eta^{\mu\nu}\, d\tau\,, \end{cases} \tag{7.3}$$

where we fixed $|\alpha| = 1$ and gauge-fixed $\varepsilon = m^{-1}$ such that τ is the proper time. Classical causality is violated, when the particle moves faster than light. In the rest frame of the particle, where $v_+^i = 0$, we can estimate this by the violation condition

$$\sum_{i=1}^d \left| \delta_{ij} \Delta X^i \, \Delta X^j \right|^2 = \sum_{i=1}^d \left| \delta_{ij} \mathrm{Re}(\Delta M^i)\, \mathrm{Re}(\Delta M^j) \right|^2 \geq c^2 \, \Delta\tau^2\,, \tag{7.4}$$

where ΔX represents the distance traveled over a time interval $\Delta\tau$. Then, using that

$$\mathrm{Re}(\Delta M^i) \sim \mathcal{N}\left(0,\, \frac{\hbar\,(1 + \cos\phi)}{2\,m}\, \Delta\tau\right)\,, \tag{7.5}$$

we find the characteristic scale for the causality violations. Causality violations are likely to occur for length scales

$$c\,\Delta\tau \lesssim \frac{\hbar\, d}{2\,m\,c}\,(1 + \cos\phi)\,, \tag{7.6}$$

where d is the dimension of space, $\phi = 0$ for Brownian motion and $\phi = \frac{\pi}{2}$ for quantum mechanics. For larger scales, the probability of such violations quickly decays to 0.

We conclude that a stochastic trajectory satisfies the stochastic notion of causality, but the sample paths may violate the stricter classical notion of causality on segments shorter than the Compton wavelength. This provides a stochastic explanation why a quantum particle cannot be localized within its own Compton wavelength.

7.3 Bell's Theorem

The stochastic theory presented in this book does not exclude the existence of an underlying (non-)deterministic hidden variable interpretation. However, since the stochastic theory reduces to quantum mechanics in the limit $\alpha = i$, Bell's theorem could provide an argument against such a hidden variable interpretation. As the Bell inequalities apply to multi-particle systems and typically involve the notion of spin, a proper analysis of Bell's theorem cannot be performed within the stochastic framework presented in this book. The stochastic theory can, however, be extended to multi-particle systems with spin. Thus, it is worth investigating whether the Bell inequalities apply to the stochastic theory.

Bell's theorem relies on the derivation of Bell inequalities for certain systems. The system considered by Bell [110, 111] is a system of two particles with spin-$\frac{1}{2}$. These particles are prepared in the singlet state

$$|s_{1,x}\, s_{2,x}\rangle = \frac{1}{2}\left(|\uparrow\downarrow\rangle - |\uparrow\downarrow\rangle\right) \tag{7.7}$$

with respect to the unit vector x. The particles are then separated, in such a way that the entanglement is preserved, and the spin of the individual particles is measured by two observers. These observers, Alice and Bob, measure the normalized spin of the particle along the unit vectors a and b, which they may choose freely. We will denote the measurement performed by Alice by s_1 and the measurement performed by Bob by s_2. Both measurements may depend on both orientations a and b, but, if we assume that the choice of orientation a cannot affect the measurement outcome at s_2 and vice versa,[4] we find

$$s_1(a, b) = s_1(a) \quad \text{and} \quad s_2(a, b) = s_2(b). \tag{7.8}$$

The outcomes of these measurements are given by

$$s_1(a) = \pm 1 \quad \text{and} \quad s_2(b) = \pm 1. \tag{7.9}$$

---

[4] In Refs. [103, 112], this is called the assumption of active locality.

Moreover, since the system is in a singlet state, we have

$$s_2(c) = -s_1(c) \quad \forall c. \tag{7.10}$$

One can calculate the expectation value of the product of two states. Quantum mechanics predicts that this expectation value is given by

$$\mathbb{E}[s_1(a) s_2(b)] = -a \cdot b. \tag{7.11}$$

On the other hand, one can derive that in a locally real theory the expectation satisfies the Bell inequality

$$\left| \mathbb{E}[s_1(a) s_2(b)] - \mathbb{E}[s_1(a) s_2(c)] \right| \le 1 + \mathbb{E}[s_1(c) s_2(b)], \tag{7.12}$$

where c is a third direction along which the spin can be measured. It is easy to find configurations for the unit vectors such that the quantum mechanics prediction violates the Bell inequality. One possible choice is the configuration that satisfies

$$a \cdot b = 0 \quad a \cdot c = \frac{\sqrt{2}}{2} \quad b \cdot c = \frac{\sqrt{2}}{2}. \tag{7.13}$$

The Bell inequality (7.12) is derived assuming that there is a hidden variable Λ that determines both the spins, such that their product can be calculated using the tower property for conditional expectations:

$$\mathbb{E}[s_1(a) s_2(b)] = \mathbb{E}[\mathbb{E}[s_1(a) s_2(b) \mid \Lambda]]$$
$$= \int \mathbb{E}[s_1(a) s_2(b) \mid \Lambda = \lambda] \, d\mu_\Lambda(\lambda)$$
$$= \int s_1(a, \lambda) s_2(b, \lambda) \, d\mu_\Lambda(\lambda), \tag{7.14}$$

where $d\mu_\Lambda(\lambda)$ is the probability distribution of the hidden variable Λ. The Bell inequality (7.12) then follows from this assumption and Eqs. (7.9) and (7.10).

The fundamental assumption made in Eq. (7.14) is that both s_1 and s_2 are Λ-measurable. However, this assumption may fail for any theory that (i) allows for the creation of entangled states and (ii) predicts an uncertainty principle. This can be seen by treating the singlet state (7.7) as a composite particle. Then, if there exists an uncertainty principle between the spin projections s_x, s_y and s_z, it immediately follows that the product $s_1(a) s_2(b)$ is only defined, if a and b are parallel. Therefore, the presence of an uncertainty relation implies that there does not exist a measure μ_Λ that determines $s_1(a) s_2(b)$ for any choice of orientations a and b.

Consequently, the failure of Eq. (7.14) can be regarded as a failure of the assumption of realism, i.e. the assumption that the product $s_1(a) s_2(b)$ is well-defined for any a and b. Alternatively, one can argue that the treatment of the entangled state as a composite object is only applicable before the separation of the two particles. Following this line of reasoning, the failure of Eq. (7.14) is due to the interplay of an uncertainty principle for the singlet state, when it is created, and a violation of locality that is implied by the existence of spatially separated entangled systems.[5]

For the stochastic theories presented in this book, (i) and (ii) hold, such that Eq. (7.14) fails. Indeed, in this framework, the position x^i and momentum operator $p_j = -\alpha\,\partial_j$ satisfy a commutation relation $[x^i, p_j] = \alpha\,\delta^i_j$. For $d = 3$, this implies a commutation relation on the angular momenta of the form $[L_i, L_j] = \alpha\,\epsilon_{ij}{}^k L_k$. Hence, once spin is included in the theory, one expects a spin-commutation relation of the form $[s_i, s_j] = \alpha\,\epsilon_{ij}{}^k s_k$. Moreover, as discussed in Sect. 7.1, one can construct entangled states by allowing for a weak violation of locality. Then, for any $\alpha \in \mathbb{C}\backslash\{0\}$, one can derive a Bell inequality, and find a detector configuration for which this inequality is violated. This statement applies to the stochastic theory irrespective of whether a hidden variable interpretation is given. In particular, for $\alpha = 1$ the stochastic theory reduces to a Brownian motion, for which the existence of a hidden variable interpretation is expected.

7.4 The Quantum Foam

As discussed in Sect. 4.3, the complex diffusion theory requires the existence of four velocity fields v_+, v_-, u_+, u_- that satisfy the constraint (4.31) in a non-relativistic theory and the constraint (5.27) in a relativistic theory. We can introduce the averages of these velocities and deviations thereof given by

$$v_\circ = \frac{1}{2}(v_+ + v_-)\,, \qquad\qquad u_\circ = \frac{1}{2}(u_+ + u_-)\,,$$
$$v_\perp = \frac{1}{2}(v_+ - v_-)\,, \qquad\qquad u_\perp = \frac{1}{2}(u_+ - u_-)\,. \tag{7.15}$$

Since $v_+(X_t, t)$ and $u_+(X_t, t)$ describe the velocity fields of X_t shortly after time t, while $v_-(X_t, t)$ and $u_-(X_t, t)$ describe the velocity fields of X_t shortly before time t, these new fields may be preferred to describe the velocity field of X at time t.

Since the fields v_+ and v_- are associated to the probability measure μ_X, they can be measured in experiment, which implies that $v_\circ$ and $v_\perp$ have a clear physical interpretation. This is not true for $u_\circ$ and $u_\perp$, but, using the constraint (4.31), $u_\perp$ can be expressed in terms of $v_\circ, v_\perp, u_\circ$, if $\phi \in (-\pi, \pi)$. Therefore, only the field $u_\circ$ lacks a physical interpretation.

If one supplements the stochastic theory with a hidden variable in the sense of Brownian motion, one obtains a physical interpretation of this velocity field $u_\circ$. In

---

[5] A similar argument is given in Refs. [103, 112]. There, it is shown that stochastic theories can violate a principle of passive locality, meaning that the observations of Alice and Bob are not conditionally independent given the prior preparation (7.7).

this physical picture, the Brownian particle is continuously interacting with many microscopic particles that induce the Brownian motion. These particles together make up a fluid and the average motion of these particles defines a velocity field for this fluid. In this interpretation, we can associate $u_\circ$ to the velocity of this fluid.

One can speculate on the nature of this background fluid with which all matter interacts. Since the fluid acts as a medium through which all matter propagates, it has a strong resemblance with the aether theory, which has been discarded as it does not respect Lorentz invariance. The aether was then replaced by the notion of spacetime in general relativity and the notion of a quantum vacuum in quantum field theory. Both spacetime and the quantum vacuum can be interpreted as a medium through which matter propagates, but, contrary to the aether theory, they are compatible with general covariance, Lorentz invariance and the gauge symmetries.

In a yet to be discovered theory of quantum gravity, it is expected that the notions of spacetime and the quantum vacuum will be replaced by the notion of a quantum foam, which is expected to be compatible with general covariance, gauge symmetries and a possibly deformed Lorentz invariance.

The fluid arising in the hidden variable interpretation of the stochastic theory is compatible with general covariance, gauge symmetries and with an Itô deformed Lorentz symmetry. Moreover, the Itô deformations vanish in the limit $\hbar \to 0$, where the theory reduces to general relativity and the limit $G \to 0$, where the theory reduces to quantum theory on flat spacetime. Therefore, this fluid can be interpreted as the quantum foam arising in various approaches to quantum gravity.[6]

We emphasize that the stochastic theory does not provide any physical laws for the microscopic behavior of the quantum foam. Instead, it describes the effective interactions between the quantum foam and all matter by means of a stochastic law. In this book, we have assumed that this law is given by Eq. (5.7), as this particular assumption allows to recover ordinary quantum theory on curved spacetime.

Theories of quantum gravity often predict Planck scale deviations from quantum theory on curved spacetime. Such deviations may be incorporated in the stochastic theory by considering more general structure relations as given in Eq. (3.16). In particular, one can consider $\beta \neq 0$, which allows the particle to make jumps at randomly distributed times. The characteristic length and time scales for these jumps are proportional to κ^{-1} and $(c\,\kappa)^{-1}$, which can be associated to the Planck length and Planck time.

---

[6] A similar observation was made by Calogero [113, 114].

Chapter 8
Discussion

Abstract This chapter briefly summarizes the main results obtained in the previous chapters, and provides an outlook on future research in the field.

8.1 Conclusion

In this book, we have reviewed and extended the theory of stochastic mechanics. In particular, we have used a generalization of the stochastic quantization procedure to describe a large class of stochastic theories. This stochastic quantization procedure can be formulated by the following postulates:

- a stochastic particle is described by a two-sided semi-martingale $X = C_\pm + \mathrm{Re}(M)$;
- the finite variation processes $C_\pm$ are such that X minimizes a stochastic action $S(X)$;
- the complex martingale $M = M_x + i\, M_y$ is such that M_x and M_y are correlated two-sided Lévy processes;
- the quadratic variation of these Lévy processes is fixed by the structure relation

$$d[M^a, M^b] = \frac{\alpha\, \hbar}{m} A^{ab}\, dt + \frac{\beta}{\kappa} B^{ab}_c\, dM^c_t \qquad (8.1)$$

with $\alpha, \beta \in \mathbb{C}$.

We have studied this theory for $\beta = 0$ and $A^{ab} = \delta^{ab}$, and shown that the resulting theory describes a general complex diffusion theory with diffusion parameter $\alpha = |\alpha|\, e^{i\phi}$. This theory reduces to a Brownian theory for $\alpha \in \mathbb{R}$, to a quantum theory for $\alpha \in i \times \mathbb{R}$ and to a classical theory for $\alpha = 0$. Moreover, the time reversal operation acts as $T(\alpha) = -\alpha$.

© The Author(s), under exclusive license to Springer Nature Switzerland AG 2023
F. Kuipers, *Stochastic Mechanics*, SpringerBriefs in Physics,
https://doi.org/10.1007/978-3-031-31448-3_8

The equivalence between the stochastic theory and a complex diffusion theory is established by the derivation of a complex diffusion equation for non-relativistic theories and a complex wave equation for relativistic theories. This derivation implies that all predictions following from these diffusion and wave equations are also predictions of the stochastic theory.

In particular, one can easily verify that the stochastic theory satisfies an energy-momentum uncertainty principle given by

$$\sigma_{x^i}\,\sigma_{p_j} \geq \frac{|\alpha|\,\hbar}{2}\left[1 + \cos(\phi)\right]\delta^i_j \,. \tag{8.2}$$

In the relativistic theory, this is supplemented by an energy momentum relation for $x^0 = c\,t$ and $E = c\,p_0$, which is given by

$$\sigma_t\,\sigma_E \geq \frac{|\alpha|\,\hbar}{2}\left[1 - \cos(\phi)\right]. \tag{8.3}$$

For $\phi = \frac{\pi}{2}$, these relations reduce to the well-known uncertainty relations from quantum mechanics. The time-energy relation can be rewritten in terms of a temperature-energy relation, by defining a temperature as a Wick rotated time:

$$T^{-1} = \pm\frac{i\,k_B\,t}{\hbar}\,. \tag{8.4}$$

This yields a temperature-energy uncertainty relation of the form

$$\sigma_{T^{-1}}\sigma_E \geq \frac{|\alpha|\,k_B}{2}\left[1 + \cos(\phi)\right], \tag{8.5}$$

which, for $\phi = 0$, reduces to the uncertainty relation encountered in statistical physics [115, 116].

Starting from the diffusion equation, one finds that, for any $\alpha \in \mathbb{C}\backslash\{0\}$, the theory can be described using an operator formalism with operators given by

$$\hat{x}^\mu = x^\mu \quad \text{and} \quad \hat{p}_\mu = -\alpha\,\hbar\,\frac{\partial}{\partial x^\mu} \tag{8.6}$$

in the position representation. This immediately yields the canonical quantization condition given by the commutation relation

$$[\hat{x}^\mu,\,\hat{p}_\nu] = \alpha\,\hbar\,\delta^\mu_\nu\,. \tag{8.7}$$

The theory can also be related to the path integral formulation by considering the characteristic functional[1]

---

[1] Alternatively, one can consider the moment generating functional $M_X(J) = \varphi_X(-i\,J)$.

$$\varphi_X(J) = \mathbb{E}\left[e^{i\int_T J_\mu(\lambda)\, X^\mu(\lambda)\, d\lambda}\right], \tag{8.8}$$

where the process X is a solution of the stochastic equations of motion. If we interpret this process as a single random variable on the path space $\mathcal{M}^T$, this expression can formally be related to the partition function $Z_X(J)$, defined by a path integral, such that

$$\varphi_X(J) = \frac{Z_X(J)}{Z_X(0)} := \frac{\int DX\, \exp\left[-\frac{S(X)}{\alpha} + i\langle J, X\rangle\right]}{\int DX\, \exp\left[-\frac{S(X)}{\alpha}\right]}, \tag{8.9}$$

where $\langle ., .\rangle$ denotes the inner product on the path space.

The stochastic theory contains two limits that are of particular interest. The first is given by $\alpha \in [0, \infty)$, which represents a Brownian motion. This limit is interesting, as it is the only diffusion theory that can be described using real martingales. In this real theory, which was discussed in Chap. 3, the probability density is related to the solution of the heat equation and is given by

$$\rho_{X_t}(x, t)\Big|_{\alpha\in(0,\infty)} = \frac{|\Psi(x, t)|}{\int |\Psi(x, t)|\, d^d x}, \tag{8.10}$$

where Ψ is a real integrable function.

For all other values of α, one requires the complex description, introduced in Chap. 4, using complex martingales and complex square integrable wave functions. For these complex theories the probability density is given by

$$\rho(x, t) = \frac{|\Psi(x, t)|^2}{\int |\Psi(x, t)|^2\, d^d x}. \tag{8.11}$$

The second interesting limit is the quantum limit, given by $\alpha \in i \times \mathbb{R}$. In this limit, the theory is unitary. This unitarity is reflected by the fact

$$\frac{d}{dt}\int |\Psi(x, t)|^2\, d^d x\Big|_{\alpha\in i\times\mathbb{R}} = 0, \tag{8.12}$$

which implies the Born rule

$$\rho(x, t)\Big|_{\alpha\in i\times\mathbb{R}} = |\Psi(x, t)|^2. \tag{8.13}$$

As shown in Chap. 5, the complex processes introduced in Chap. 4 also provide a natural framework for the study of relativistic stochastic processes. This relativistic theory provides a consistent framework for the study of relativistic Brownian motion, but also for the study of a single relativistic quantum particle and anything in between.

The fact that the relativistic stochastic theory is well defined for the single particle is particularly interesting, as the standard description of relativistic quantum particles requires an extension to quantum field theory. In the stochastic theory, such an extension is not necessary, since the norm is defined through an expectation value. As discussed in Sect. 7.2, the sample paths of the process may go off-shell and may even violate the classical notion of causality for short time intervals, but the stochastic notion of causality is obeyed at all times, which ensures that the sign of the norm is preserved in the theory.[2]

In Chap. 6, we extended the complex stochastic theory to Riemannian and Lorentzian geometry. This extension was performed using the framework of second order geometry, which provides an elegant way to construct covariant Itô integrals. Using this framework, we found that the usual spacetime symmetries are deformed to the Itô deformed Galilean symmetry in non-relativistic theories and the Itô deformed Poincaré symmetry in relativistic theories. Moreover, we noticed that these deformations vanish in the classical limit $\hbar \to 0$ and the no-gravity limit $G \to 0$.

The Itô deformations only occur when the theory is evaluated in the Itô framework. In a Stratonovich framework, the Galilean group and Poincaré group remain to be the correct symmetries of the stochastic theory in curved space(time). Therefore, there may exist equivalent formulations of a quantum theory in curved spacetime: one that respects the classical spacetime symmetries and one that Itô deforms the classical spacetime symmetries.

Canonical formulations of quantum theories respect the classical spacetime symmetries, as they are typically constructed in a first order formalism. However, this is not true for path integrals. As these are second order objects, the path integral measure is not covariant. Thus, for path integrals, covariance can be regained by Itô deforming the Galilean and Poincaré symmetries.

We emphasize that these Itô deformations of the spacetime symmetries follow from imposing that ordinary quantum theories on curved space(time) must respect general covariance. Other deformations of spacetime symmetries have been suggested in the literature, cf. e.g. [117], but such deformations are different form the Itô deformations, as they typically require the introduction of a minimal length scale, which is inspired by quantum gravity.

8.2 Outlook

The complex stochastic theory unifies quantum mechanics with Brownian motion in a single mathematical framework. The presented theory is, however, limited to the dynamics of a single (non-)relativistic particle subjected to a scalar and vector potential on a pseudo-Riemannian manifold. An important next step is the extension of this framework to multi-particle systems, field theories and to particles with spin.

---

[2] When compared to a field theoretic formulation of relativistic quantum theories, the violation of classical causality is associated to the presence of eigenstates with negative eigenenergy, while the condition of stochastic causality is associated to the condition that commutators vanish outside the lightcone.

Various examples of such extensions have been studied in the literature, cf. e.g. [22, 28, 36, 53, 57, 58, 61, 72–84], but no general framework exists yet. The complex stochastic theory, as presented in this book, allows for a systematic study of such extensions.

Of particular interest is the analysis of the Itô deformations in a field theoretic framework. Given the Klein-Gordon equation that was derived for the single particle in Sect. 6.4, it is expected that the Pauli-DeWitt term must be present in the quantum action of any scalar field. This implies that a minimally coupled classical scalar theory in $3 + 1$ dimensions will become conformally coupled, when the theory is quantized. As we have not discussed spin, it is unclear whether further corrections can be expected for fermions or vector fields. In general, any such correction to the Lagrangian will predict deviations from the standard models of particle physics and cosmology in the regime where neither quantum mechanics nor gravity can be neglected.

Further corrections are to be expected from the complexification of the theory. In the complexified theory, the Lagrangian depends on the real position X and the complex velocity W. In the Stratonovich framework, this Lagrangian is given by

$$L^\circ(x, w_\circ, \varepsilon) = \frac{1}{2\varepsilon} g_{\mu\nu}(x) w_\circ^\mu w_\circ^\nu - \frac{\varepsilon m^2}{2} + q A_\mu(x) w_\circ^\mu. \qquad (8.14)$$

Hence, the complexification introduces new terms, related to the velocity $u_\circ$, given by

$$L_q^\circ(x, v_\circ, u_\circ, \varepsilon) = -\frac{1}{2\varepsilon} g_{\mu\nu}(x) u_\circ^\mu u_\circ^\nu + \frac{i}{\varepsilon} g_{\mu\nu}(x) v_\circ^\mu u_\circ^\nu + i q A_\mu(x) u_\circ^\mu. \qquad (8.15)$$

Since these corrections involve the complex velocity $u_\circ$, they can be associated to the velocity of the background field or quantum foam, cf. Sect. 7.4. Therefore, it is expected that, in a field theoretic framework, the stochastic theory predicts the existence of a background field that contributes to the energy balance of the universe.

As discussed in Sect. 7.4, the stochastic theory is a natural framework for the study of a fluctuating spacetime, which appears in various theories of quantum gravity. However, the interpretation slightly differs from most approaches to quantum gravity. According to the stochastic theory, all quantum effects are induced by the stochastic fluctuations of the spacetime foam. The stochastic theory does not predict any Planck scale corrections to ordinary quantum physics. However, various classes of such corrections can be incorporated by studying the general structure relation (8.1) with $\beta \neq 0$. Stochastic theories defined by such a general structure relation can be studied as a model for quantum gravity and may be related to other approaches to quantum gravity.

In the stochastic theory, gravity has itself a rather special status, due to the fact that the variation of the action with respect to a process X induces a variation with respect to both the velocity $V_\pm$ and the second order velocity V_2. This second order velocity is proportional to the metric g. Therefore, a variation with respect to the metric is implied by the variation with respect to any other field. This verifies that gravity couples to all matter, but also suggests that gravity is an emergent force. In this picture, gravity emerges as the field must minimize its action with respect to both its classical trajectory and its quantum fluctuations. The latter may be achieved by deforming spacetime, thus by inducing gravity.

Various of these extensions of the stochastic theory are subject of current investigation. More generally, the stochastic theory, as presented in this book, deepens the connection between quantum mechanics and stochastic theories. This might be used in various ways: it allows to further exploit the tools from stochastic analysis to study quantum systems, while, on the other hand, it allows for further exploitation of the tools from quantum theory to study stochastic processes in applications ranging from soft condensed matter theory to finance.

Appendix A
Review of Probability Theory

Abstract Appendices A and B review the relevant mathematical terminology from measure theoretic probability theory, which is used throughout the book. Appendix C discusses elementary notions from stochastic calculus. Appendix D shows how stochastic calculus can be extended to the context of pseudo-Riemannian geometry using second order geometry. Appendix E derives the Itô Lagrangian, which is the stochastic generalization of the classical Lagrangian. Appendix F shows how the stochastic Stratonovich-Euler-Lagrange, the stochastic Itô-Euler-Lagrange and the stochastic Hamilton-Jacobi-Bellman equations can be derived from a stationary action principle using stochastic variational calculus.

In this appendix, we list various standard definitions and results from probability theory. For a more detailed discussion of these definitions and results, we refer to textbooks on probability theory and stochastic processes, e.g. Refs. [70, 118–121].

A.1 Probability Spaces

The most elementary object in probability theory is a probability space. This is a sample space Ω, i.e. a collection of the possible outcomes in a probability experiment, together with the probability of occurrence for each outcome $\omega \in \Omega$. Crucially, these probabilities cannot be defined on the outcomes $\omega \in \Omega$ themselves. Instead, they must be defined on events in a sigma algebra over Ω.

Definition (*Sigma algebra*) Given a set Ω, a sigma algebra over Ω is a set $\Sigma(\Omega) = \{A \mid A \subseteq \Omega\}$ satisfying the following properties

- $\Omega \in \Sigma$;
- $\forall A \in \Sigma, \quad \Omega \setminus A \in \Sigma$;
- $\forall A_1, A_2, \ldots \in \Omega, \quad A_1 \cup A_2 \cup \ldots \in \Sigma$.

Definition (*Borel sigma algebra*) Given a topological space $(\mathcal{M}, \mathfrak{T})$, a Borel set is a set that can be obtained by taking countable unions, countable intersections and complements of the sets in the topology $\mathfrak{T}$. The collection of all Borel sets is the Borel sigma algebra $\mathcal{B}(\mathcal{M})$.

© The Editor(s) (if applicable) and The Author(s), under exclusive license to Springer
Nature Switzerland AG 2023
F. Kuipers, *Stochastic Mechanics*, SpringerBriefs in Physics,
https://doi.org/10.1007/978-3-031-31448-3

Definition (*Measurable space*) Given a set Ω, a measurable space is a tuple (Ω, Σ), where $\Sigma = \Sigma(\Omega)$ is a sigma algebra over Ω.

Definition (*Probability measure*) Given a measurable space (Ω, Σ), a probability measure is a function $\mathbb{P} : \Sigma \to [0, 1]$, such that

- $\mathbb{P}(\Omega) = 1$,
- $\mathbb{P}(\bigcup_i A_i) = \sum_i \mathbb{P}(A_i)$ for any countable collection $\{A_i \in \Sigma \mid i \in \{1, 2, ...\}\}$ of pairwise disjoint sets.

Definition (*Probability space*) A probability space is a tuple $(\Omega, \Sigma, \mathbb{P})$, where (Ω, Σ) is a measurable space and $\mathbb{P} : \Sigma \to [0, 1]$ is a probability measure.

Definition (*Almost surely*) Given a probability space $(\Omega, \Sigma, \mathbb{P})$, we say that an event $A \in \Sigma$ occurs almost surely (abbreviated to a.s.), if $\mathbb{P}(A) = 1$.

A.2 Random Variables

Random variables are the main object of study in probability theory. Essentially, these are functions that translate the outcomes of a probability experiment into probabilistic events in the real world.

Definition (*Measurable function*) Given two measurable spaces (Ω_1, Σ_1) and (Ω_2, Σ_2), a function $X : (\Omega_1, \Sigma_1) \to (\Omega_2, \Sigma_2)$ is a measurable function, if

- $\forall U \in \Sigma_2, \quad X^{-1}(U) := \{\omega \in \Omega \mid X(\omega) \in U\} \in \Sigma_1$.

Definition (*Borel measurable function*) Given two topological spaces $\mathcal{M}$ and $\mathcal{N}$, a function $f : \mathcal{M} \to \mathcal{N}$ is called Borel measurable, if

- $\forall U \in \mathcal{B}(\mathcal{N}), \quad f^{-1}(U) := \{x \in \mathcal{M} \mid f(x) \in U\} \in \mathcal{B}(\mathcal{M})$.

Definition (*Random variable*) Given a probability space $(\Omega, \Sigma, \mathbb{P})$ and a measurable space $(\mathcal{M}, \mathcal{B}(\mathcal{M}))$, an $\mathcal{M}$-valued random variable is a measurable function $X : (\Omega, \Sigma, \mathbb{P}) \to (\mathcal{M}, \mathcal{B}(\mathcal{M}))$.

Definition (*Real random variable*) Given a probability space $(\Omega, \Sigma, \mathbb{P})$ and the real space $\mathbb{R}^n$, an n-dimensional real random variable is a random variable $X : (\Omega, \Sigma, \mathbb{P}) \to (\mathbb{R}^n, \mathcal{B}(\mathbb{R}^n))$.

Definition (*Complex random variable*) Given a probability space $(\Omega, \Sigma, \mathbb{P})$ and the complex space $\mathbb{C}^n$, an n-dimensional complex random variable is a random variable $Z : (\Omega, \Sigma, \mathbb{P}) \to (\mathbb{C}^n, \mathcal{B}(\mathbb{C}^n))$, such that $Z = X + \mathrm{i}\, Y$, where $X, Y : (\Omega, \Sigma, \mathbb{P}) \to (\mathbb{R}^n, \mathcal{B}(\mathbb{R}^n))$ are real random variables.

Definition (*Distribution of a random variable*) A random variable $X : (\Omega, \Sigma, \mathbb{P}) \to (\mathcal{M}, \mathcal{B}(\mathcal{M}))$ induces a probability measure $\mu_X = \mathbb{P} \circ X^{-1}$ on $(\mathcal{M}, \mathcal{B}(\mathcal{M}))$. The induced function $d\mu_X : \mathcal{M} \to [0, 1]$ is called the distribution of X.

Theorem (Functions of random variables) *Given a random variable $X : (\Omega, \Sigma, \mathbb{P})$ $\rightarrow (\mathcal{M}, \mathcal{B}(\mathcal{M}))$ and a Borel measurable function f defined on $\mathcal{M}$, the composition $f \circ X$ is a random variable.*

A.3 Expectation Value

In the study of random variables, one is often interested in their expectation values. The expectation value of a random variable is defined by a Lebesgue integral.

Definition (*Lebesgue integrability*) A real random variable $X : (\Omega, \Sigma, \mathbb{P}) \rightarrow$ $(\mathbb{R}, \mathcal{B}(\mathbb{R}))$ is integrable, if and only if the Lebesgue integral $\int_\Omega |X(\omega)| \, d\mathbb{P}(\omega)$ converges.

More generally, given a smooth manifold $\mathcal{M}$, a random variable $X : (\Omega, \Sigma, \mathbb{P}) \rightarrow$ $(\mathcal{M}, \mathcal{B}(\mathcal{M}))$ is integrable, if and only if the Lebesgue integral $\int_\Omega |f[X(\omega)]| \, d\mathbb{P}(\omega)$ converges for all $f \in C^\infty(\mathcal{M})$.

Definition (*Square integrability*) A real random variable $X : (\Omega, \Sigma, \mathbb{P}) \rightarrow$ $(\mathbb{R}, \mathcal{B}(\mathbb{R}))$ is square integrable, if and only if the Lebesgue integral $\int_\Omega |X(\omega)|^2 \, d\mathbb{P}(\omega)$ converges.

More generally, given a smooth manifold $\mathcal{M}$, a random variable $X : (\Omega, \Sigma, \mathbb{P}) \rightarrow$ $(\mathcal{M}, \mathcal{B}(\mathcal{M}))$ is square integrable, if and only if the Lebesgue integral $\int_\Omega |f[X(\omega)]|^2$ $d\mathbb{P}(\omega)$ converges for all $f \in C^\infty(\mathcal{M})$.

Definition (*Expectation value*) Given an integrable random variable $X : (\Omega, \Sigma, \mathbb{P})$ $\rightarrow (\mathbb{R}^n, \mathcal{B}(\mathbb{R}^n))$, the expectation value of the i-th component X^i is given by the Lebesgue integral

$$\mathbb{E}[X^i] = \int_\Omega X^i(\omega) \, d\mathbb{P}(\omega) \,. \tag{A.1}$$

More generally, given an integrable random variable $X : (\Omega, \Sigma, \mathbb{P}) \rightarrow (\mathcal{M}, \mathcal{B}(\mathcal{M}))$ and a function $f \in C^\infty(\mathcal{M})$, the expectation value of $f(X)$ is given by the Lebesgue integral

$$\mathbb{E}[f(X)] = \int_\Omega f[X(\omega)] \, d\mathbb{P}(\omega) \,. \tag{A.2}$$

Theorem (Expectation value I) *Given a probability space $(\Omega, \Sigma, \mathbb{P})$, a measurable space $(\mathcal{M}, \mathcal{B}(\mathcal{M}))$, a random variable $X : \Omega \rightarrow \mathcal{M}$ and a Borel measurable function $f : \mathcal{M} \rightarrow \mathbb{R}$ such that $f \circ X$ is integrable, then*

$$\mathbb{E}[f(X)] = \int_\mathcal{M} f(x) \, d\mu_X(x) \,, \tag{A.3}$$

where $d\mu_X$ is the distribution of X on $\mathcal{M}$.

Definition (*Probability density*) Given a probability space $(\Omega, \Sigma, \mathbb{P})$, an n-dimensional pseudo-Riemannian manifold $(\mathcal{M}, g)$ and an integrable random variable $X : \Omega \to \mathcal{M}$, a Borel measurable function $\rho_X : \mathcal{M} \to [0, \infty)$ is a probability density for X, if

$$\mu_X(B) = \int_B \rho_X(x) \, dV_R(x) = \int_B \sqrt{|g(x)|} \, \rho_X(x) \, d^n x \quad \forall \, B \in \mathcal{B}(\mathcal{M}), \quad (A.4)$$

where V_R is the Riemann measure on $(\mathcal{M}, g)$.

Theorem (Expectation value II) *Given a probability space $(\Omega, \Sigma, \mathbb{P})$, an n-dimensional pseudo-Riemannian manifold $(\mathcal{M}, g)$, a random variable $X : \Omega \to \mathcal{M}$ with probability density ρ_X and a Borel measurable function $f : \mathcal{M} \to \mathbb{R}$, such that $f \circ X$ is integrable, then*

$$\mathbb{E}[f(X)] = \int_{\mathcal{M}} \sqrt{|g(x)|} \, f(x) \, \rho_X(x) \, d^n x . \quad (A.5)$$

A.4 Conditional Expectation

In practice, one is often interested in properties of a random variable, while one is already given some information. In order to study random variables, when such partial information is given, we require the notion of conditional expectation.

Definition (*Sigma algebra generated by a random variable*) Given a random variable $X : (\Omega, \Sigma, \mathbb{P}) \to (\mathcal{M}, \mathcal{B}(\mathcal{M}))$, the sigma algebra generated by X is given by

$$\sigma(X) = \big\{ \{\omega \in \Omega \mid X(\omega) \in B\} \, \big| \, B \in \mathcal{B}(\mathcal{M}) \big\} . \quad (A.6)$$

Definition (*$\mathcal{G}$-measurability*) Given a random variable $X : (\Omega, \Sigma, \mathbb{P}) \to (\mathcal{M}, \mathcal{B}(\mathcal{M}))$ and a sigma algebra $\mathcal{G}(\Omega)$, X is $\mathcal{G}$-measurable, if $\sigma(X) \subseteq \mathcal{G}$.

Definition (*Stochastic independence*) Given a probability space $(\Omega, \Sigma, \mathbb{P})$ and two sub sigma algebras $\mathcal{G}_1, \mathcal{G}_2 \subseteq \Sigma$, we call $\mathcal{G}_1$ and $\mathcal{G}_2$ independent, denoted by $\mathcal{G}_1 \perp\!\!\!\perp \mathcal{G}_2$, if

$$\mathbb{P}(A \cap B) = \mathbb{P}(A) \mathbb{P}(B) \quad \forall \, A \in \mathcal{G}_1, \, B \in \mathcal{G}_2 . \quad (A.7)$$

Definition (*Independent random variables*) Given a probability space $(\Omega, \Sigma, \mathbb{P})$ and two random variables X, Y defined on this space, we say that X and Y are independent, denoted by $X \perp\!\!\!\perp Y$, if $\sigma(X)$ and $\sigma(Y)$ are independent sigma algebras.

Theorem (Functions of independent random variables) *Given two independent random variables $X, Y : (\Omega, \Sigma, \mathbb{P}) \to (\mathcal{M}, \mathcal{B}(\mathcal{M}))$, and two Borel measurable functions f, g defined on $\mathcal{M}$, then $f(X) \perp\!\!\!\perp g(Y)$.*

Definition (*Conditional expectation*) Given an integrable random variable X : $(\Omega, \Sigma, \mathbb{P}) \to (\mathbb{R}, \mathcal{B}(\mathbb{R}))$ and a sub sigma algebra $\mathcal{G} \subseteq \Sigma$, the conditional expectation of X given $\mathcal{G}$, denoted by $\mathbb{E}[X \mid \mathcal{G}]$, is any real random variable that is $\mathcal{G}$-measurable and satisfies

$$\int_A \mathbb{E}[X \mid \mathcal{G}](\omega) \, d\mathbb{P}(\omega) = \int_A X(\omega) \, d\mathbb{P}(\omega) \quad \forall \, A \in \mathcal{G}. \tag{A.8}$$

If $\mathcal{G} = \sigma(Y)$ for some other random variable Y, we write $\mathbb{E}[X \mid Y] := \mathbb{E}[X \mid \sigma(Y)]$.

Theorem (Properties of conditional expectation) *Given two integrable random variables $X, Y : (\Omega, \Sigma, \mathbb{P}) \to (\mathbb{R}, \mathcal{B}(\mathbb{R}))$ and a sub sigma algebra $\mathcal{G} \subseteq \Sigma$, the conditional expectation satisfies the following properties:*

- *(Linearity):* $\mathbb{E}[a \, X + b \, Y \mid \mathcal{G}] = a \, \mathbb{E}[X \mid \mathcal{G}] + b \, \mathbb{E}[Y \mid \mathcal{G}]$ *for any $a, b \in \mathbb{R}$;*
- *(Taking out what is known):* *If X is $\mathcal{G}$-measurable, then* $\mathbb{E}[X \, Y \mid \mathcal{G}] = X \, \mathbb{E}[Y \mid \mathcal{G}]$;
- *(Tower property):* $\mathbb{E}[\, \mathbb{E}[X \mid \mathcal{G}] \mid \mathcal{H}] = \mathbb{E}[X \mid \mathcal{H}]$ *for any sub sigma algebra $\mathcal{H} \subset \mathcal{G}$;*
- *(Independence):* *If $X \perp\!\!\!\perp \mathcal{G}$, then* $\mathbb{E}[X \mid \mathcal{G}] = \mathbb{E}[X]$;
- *(Jensen's inequality):* $\phi(\mathbb{E}[X \mid \mathcal{G}]) \leq \mathbb{E}[\phi(X) \mid \mathcal{G}]$ *for any convex function ϕ.*

A.5 Change of Measure

The measure on a measurable space is not uniquely defined, but there exist relations between different measures on the same measurable space. These relations are expressed by the Radon-Nykodým theorem.

Theorem (Change of measure) *Given a probability space $(\mathcal{M}, \mathcal{B}(\mathcal{M}), \mu)$ and a Borel measurable function $h : \mathcal{M} \to [0, \infty)$, such that $\int_{\mathcal{M}} h(x) \, d\mu(x) = 1$, the object defined by*

$$\tilde{\mu}(A) = \int_B h(x) \, d\mu(x) \quad \forall \, B \in \mathcal{B}(\mathcal{M}) \tag{A.9}$$

is a probability measure on $(\mathcal{M}, \mathcal{B}(\mathcal{M}))$. Furthermore, for every Borel measurable function $f : \mathcal{M} \to [0, \infty)$,

$$\int_{\mathcal{M}} f(x) \, d\tilde{\mu}(x) = \int_{\mathcal{M}} f(x) \, h(x) \, d\mu(x). \tag{A.10}$$

Moreover, if $h : \mathcal{M} \to (0, \infty)$, then

$$\int_{\mathcal{M}} g(x) \, d\mu(x) = \int_{\mathcal{M}} \frac{g(x)}{h(x)} \, d\tilde{\mu}(x) \tag{A.11}$$

for every Borel measurable function $g : \mathcal{M} \to [0, \infty)$.

Definition (*Absolutely continuous*) Given a measurable space $(\mathcal{M}, \mathcal{B}(\mathcal{M}))$ and two measures μ and $\tilde{\mu}$ defined on this space, the measure $\tilde{\mu}$ is said to be absolutely continuous with respect to μ, if

$$\tilde{\mu}(B) = 0 \quad \forall B \in \{A \in \mathcal{B}(\mathcal{M}) \mid \mu(A) = 0\}. \tag{A.12}$$

Theorem (Radon-Nykodým) *Given a measurable space $(\mathcal{M}, \mathcal{B}(\mathcal{M}))$ and two probability measures μ and $\tilde{\mu}$ defined on this space, such that $\tilde{\mu}$ is absolutely continuous with respect to μ, then there exists a Borel measurable function $h : \mathcal{M} \to [0, \infty)$, such that*

- $\int_{\mathcal{M}} h(x) \, d\mu(x) = 1$;
- $\tilde{\mu}(B) = \int_{B} h(x) \, d\mu(x) \quad \forall B \in \mathcal{B}(\mathcal{M})$.

This function is denoted by

$$h = \frac{d\tilde{\mu}}{d\mu} \tag{A.13}$$

and called the Radon-Nykodým derivative of $\tilde{\mu}$ with respect to μ.

Corollary (Probability density) *Given a pseudo-Riemannian manifold $\mathcal{M}$, a random variable $X : (\Omega, \Sigma, \mathbb{P}) \to (\mathcal{M}, \mathcal{B}(\mathcal{M}))$, such that $\mu_X = \mathbb{P} \circ X^{-1}$ is absolutely continuous with respect to the Riemann measure V_R on $(\mathcal{M}, \mathcal{B}(\mathcal{M}))$, then X has a probability density ρ_X, which is given by the Radon-Nykodým derivative*

$$\rho_X = \frac{d\mu_X}{dV_R}. \tag{A.14}$$

A.6 L^2-spaces

In order to do analysis and to study convergence of random variables, one requires a notion of distance. This is provided by the L^p-norm. Here, we focus on the case $p = 2$.

Definition (L^2-*norm*) Given a random variable $X : (\Omega, \Sigma, \mathbb{P}) \to (\mathbb{K}^n, \mathcal{B}(\mathbb{K}^n))$ with $\mathbb{K} \in \{\mathbb{R}, \mathbb{C}\}$, the L^2-norm for X is defined by

$$||X|| = \sqrt{\mathbb{E}\left[\delta_{ij} \overline{X^i} X^j\right]}. \tag{A.15}$$

Definition (L^2-*space*) Given a probability space $(\Omega, \Sigma, \mathbb{P})$, the spaces

$$L^2(\Omega, \Sigma, \mathbb{P}) = \left\{ X : (\Omega, \Sigma, \mathbb{P}) \to (\mathbb{K}^n, \mathcal{B}(\mathbb{K}^n)) \, \middle| \, ||X|| < \infty \right\} \tag{A.16}$$

with $\mathbb{K} \in \{\mathbb{R}, \mathbb{C}\}$ are the real and complex L^2-space over $(\Omega, \Sigma, \mathbb{P})$.

Definition (*Stochastic Riemannian distance*) Given a Riemannian manifold $(\mathcal{M}, g)$ and random variables $X, Y : (\Omega, \Sigma, \mathbb{P}) \to (\mathcal{M}, \mathcal{B}(\mathcal{M}))$, for any $\omega \in \Omega$ we denote the set of piecewise smooth curves between $X(\omega)$ and $Y(\omega)$ by

$$\mathcal{C}_{X,Y}(\omega) = \left\{ \gamma(\omega) : [0, T] \to \mathcal{M} \,\middle|\, \gamma_0(\omega) = X(\omega),\ \gamma_T(\omega) = Y(\omega) \right\}. \quad (A.17)$$

We define a stochastic Riemannian distance function on the space of all square integrable random variables $X : \Omega \to \mathcal{M}$ by

$$d(X, Y)(\omega) = \begin{cases} \inf \left\{ \int_0^T \sqrt{g(\dot{\gamma}_t, \dot{\gamma}_t)(\gamma_t)} \, dt \,\middle|\, \gamma \in \mathcal{C}_{X,Y}(\omega) \right\} & \text{if } \mathcal{C}_{X,Y}(\omega) \neq \emptyset, \\ \infty & \text{if } \mathcal{C}_{X,Y}(\omega) = \emptyset. \end{cases}$$
$$(A.18)$$

Definition (*Stochastic Lorentzian distance*) Given a Lorentzian manifold $(\mathcal{M}, g)$ with $(- + ... +)$ signature and random variables $X, Y : (\Omega, \Sigma, \mathbb{P}) \to (\mathcal{M}, \mathcal{B}(\mathcal{M}))$, for any $\omega \in \Omega$ we denote the set of causal and anti-causal piecewise smooth curves between $X(\omega)$ and $Y(\omega)$ by respectively

$$\mathcal{C}^{ca}_{X,Y}(\omega) = \left\{ \gamma(\omega) : [0, T] \to \mathcal{M} \,\middle|\, \gamma_0(\omega) = X(\omega),\ \gamma_T(\omega) = Y(\omega),\ g(\dot{\gamma}_t, \dot{\gamma}_t)(\gamma_t)(\omega) \leq 0 \,\forall t \in [0, T] \right\}$$
$$\mathcal{C}^{ac}_{X,Y}(\omega) = \left\{ \gamma(\omega) : [0, T] \to \mathcal{M} \,\middle|\, \gamma_0(\omega) = X(\omega),\ \gamma_T(\omega) = Y(\omega),\ g(\dot{\gamma}_t, \dot{\gamma}_t)(\gamma_t)(\omega) \geq 0 \,\forall t \in [0, T] \right\}$$
$$(A.19)$$

We define stochastic Lorentzian distance functions on the space of all square integrable random variables $X : \Omega \to \mathcal{M}$ by

$$\Delta\tau(X, Y)(\omega) = \begin{cases} \sup \left\{ \int_0^T \sqrt{-g(\dot{\gamma}_t, \dot{\gamma}_t)(\gamma_t)} \, dt \,\middle|\, \gamma \in \mathcal{C}^{ca}_{X,Y}(\omega) \right\} & \text{if } \mathcal{C}^{ca}_{X,Y}(\omega) \neq \emptyset, \\ -\infty & \text{if } \mathcal{C}^{ca}_{X,Y}(\omega) = \emptyset. \end{cases}$$
$$(A.20)$$

$$\Delta s(X, Y)(\omega) = \begin{cases} \sup \left\{ \int_0^T \sqrt{g(\dot{\gamma}_t, \dot{\gamma}_t)(\gamma_t)} \, dt \,\middle|\, \gamma \in \mathcal{C}^{ac}_{X,Y}(\omega) \right\} & \text{if } \mathcal{C}^{ac}_{X,Y}(\omega) \neq \emptyset, \\ -\infty & \text{if } \mathcal{C}^{ac}_{X,Y}(\omega) = \emptyset, \end{cases}$$
$$(A.21)$$

which we call the proper time and proper distance between X and Y.

Definition (*Induced L^2-norm*) Given a random variable $X : (\Omega, \Sigma, \mathbb{P}) \to (\mathcal{M}, \mathcal{B}(\mathcal{M}))$ and a Borel measurable function $f : \mathcal{M} \to \mathbb{K}$ with $\mathbb{K} \in \{\mathbb{R}, \mathbb{C}\}$, the L^2-norm for $f \circ X$ is defined by

$$\|f(X)\| = \sqrt{\mathbb{E}\big[|f(X)|^2\big]}. \quad (A.22)$$

Definition (*Induced L^2-space*) Given a random variable $X : (\Omega, \Sigma, \mathbb{P}) \to (\mathcal{M}, \mathcal{B}(\mathcal{M}))$, the spaces

$$L^2(\mathcal{M}, \mathcal{B}(\mathcal{M}), \mu_X) = \left\{ f : (\mathcal{M}, \mathcal{B}(\mathcal{M})) \to (\mathbb{K}, \mathcal{B}(\mathbb{K})) \,\middle|\, ||f(X)|| < \infty \right\} \quad \text{(A.23)}$$

with $\mathbb{K} \in \{\mathbb{R}, \mathbb{C}\}$ are the real and complex L^2-space over $(\mathcal{M}, \mathcal{B}(\mathcal{M}), \mu_X)$.

Theorem (*Properties of L^2-spaces*) *L^2-spaces have the following properties:*

- L^2 *is a Hilbert space with inner product*
  - $X \cdot Y = \mathbb{E}[\delta_{ij} \overline{X}^i Y^j]$ *on $L^2(\Omega, \Sigma, \mathbb{P})$,*
  - $\langle f, g \rangle_{\mu_X} = \int_{\mathcal{M}} \overline{f(x)}\, g(x)\, d\mu_X(x)$ *on $L^2(\mathcal{M}, \mathcal{B}(\mathcal{M}), \mu_X)$;*
- L^2 *is self-dual, i.e. $L^2 \cong (L^2)^*$.*

A.7 Generating Functions

Definition (*Dual norm*) Given the space $L^2(\Omega, \Sigma, \mathbb{P})$, the norm on the dual space $[L^2(\Omega, \Sigma, \mathbb{P})]^*$ is given by

$$||a|| = \sup \left\{ \mathbb{E}[|a(X)|] \,\middle|\, X \in L^2(\Omega, \Sigma, \mathbb{P}), ||X|| \leq 1 \right\}. \quad \text{(A.24)}$$

Definition (*Moment generating function*) Given a real random variable $X : (\Omega, \Sigma, \mathbb{P}) \to (\mathbb{R}^n, \mathcal{B}(\mathbb{R}^n))$ and a linear form $a \in [L^2(\Omega, \Sigma, \mathbb{P})]^*$, the moment generating function of X is defined by

$$M_X(a) := \mathbb{E}\left[e^{a(X)} \right], \quad \text{(A.25)}$$

where $||a|| < \rho$ with ρ the radius of convergence of M_X.

Given a complex random variable $Z : (\Omega, \Sigma, \mathbb{P}) \to (\mathbb{C}^n, \mathcal{B}(\mathbb{C}^n))$ and a linear form $a \in [L^2(\Omega, \Sigma, \mathbb{P})]^*$, the moment generating function of Z is defined by

$$M_Z(a) := \mathbb{E}\left[e^{\overline{a}(X)} \right], \quad \text{(A.26)}$$

where $||a|| < \rho$ with ρ the radius of convergence of M_Z.

These functions are called the moment generating functions, since they generate the moments of random variables. Indeed, it can easily be verified that

$$\mathbb{E}\left[\prod_{i=1}^{n} (X^i)^{k_i} \right] = \prod_{i=1}^{n} \frac{d^{k_i}}{da_i^{k_i}} \mathbb{E}\left[e^{a_j X^j} \right]\Big|_{a=0} \quad \forall k \in \hat{\mathbb{N}}^n,$$

$$\mathbb{E}\left[\prod_{i=1}^{n} (Z^i)^{k_i} \right] = \prod_{i=1}^{n} \frac{d^{k_i}}{d\overline{a}_i^{k_i}} \mathbb{E}\left[e^{\overline{a}_j Z^j} \right]\Big|_{a=0} \quad \forall k \in \hat{\mathbb{N}}^n. \quad \text{(A.27)}$$

Definition (*Characteristic function*) Given a real random variable $X : (\Omega, \Sigma, \mathbb{P}) \rightarrow (\mathbb{R}^n, \mathcal{B}(\mathbb{R}^n))$ and a linear form $a \in [L^2(\Omega, \Sigma, \mathbb{P})]^*$, the characteristic function of X is defined by

$$\varphi_X(a) := \mathbb{E}\left[e^{i\,a(X)}\right]. \tag{A.28}$$

Given a complex random variable $Z : (\Omega, \Sigma, \mathbb{P}) \rightarrow (\mathbb{C}^n, \mathcal{B}(\mathbb{C}^n))$ and a linear form $a \in [L^2(\Omega, \Sigma, \mathbb{P})]^*$, the characteristic function of Z is defined by

$$\varphi_Z(a) := \mathbb{E}\left[e^{i\,\mathrm{Re}[\overline{a}(Z)]}\right]. \tag{A.29}$$

If X admits a probability density, then the characteristic function φ_X and probability density ρ_X are each others Fourier transform. Similarly, if $Z = X + i\,Y$ admits a probability density, then the characteristic function φ_Z and probability density $\rho_Z = \rho_{(X,Y)}$ are each others Fourier transform.

Appendix B
Review of Stochastic Processes

In this appendix, we list various standard definitions and results from the theory of stochastic processes. For a more detailed discussion of these definitions and results, we refer to textbooks on probability theory and stochastic processes, e.g. Refs. [70, 118–121].

B.1 Stochastic Processes

Definition (*Stochastic process*) Given a probability space $(\Omega, \Sigma, \mathbb{P})$, a measurable space $(\mathcal{M}, \mathcal{B}(\mathcal{M}))$ and a set $\mathcal{T} \subseteq \mathbb{R}$, an $\mathcal{M}$-valued stochastic process is a family of random variables $\{X_t \mid t \in \mathcal{T}\}$ with $X_t : (\Omega, \Sigma, \mathbb{P}) \to (\mathcal{M}, \mathcal{B}(\mathcal{M})) \quad \forall\, t \in \mathcal{T}$.

The possible outcomes of the stochastic process are given by the sample paths

Definition (*Sample path*) Given a stochastic process $X : \mathcal{T} \times (\Omega, \Sigma, \mathbb{P}) \to (\mathcal{M}, \mathcal{B}(\mathcal{M}))$ and an event $\omega \in \Omega$, a sample path is a function $X(\cdot, \omega) : \mathcal{T} \to \mathcal{M}$.

B.2 Conditioning

One is often interested in properties of a stochastic process at a time t, given that certain events occur in the past or future. In order to study such properties, we require an object that collects information about past or future events. This object is called a filtration.

Definition (*Filtration*) Given a measurable space $(\Omega, \Sigma, \mathbb{P})$ and a set $\mathcal{T} \subseteq \mathbb{R}$, a filtration (or past filtration) is a family $\vec{\mathcal{F}} = \{\vec{\mathcal{F}}_t \mid t \in \mathcal{T}\}$ such that $\{\emptyset, \Omega\} \subseteq \vec{\mathcal{F}}_s \subseteq \vec{\mathcal{F}}_t \subseteq \Sigma \quad \forall s < t \in \mathcal{T}$.

© The Editor(s) (if applicable) and The Author(s), under exclusive license to Springer Nature Switzerland AG 2023
F. Kuipers, *Stochastic Mechanics*, SpringerBriefs in Physics,
https://doi.org/10.1007/978-3-031-31448-3

Definition (*Time reversed filtration*) Given a measurable space $(\Omega, \Sigma, \mathbb{P})$ and a set $T \subseteq \mathbb{R}$, a time reversed filtration (or future filtration) is a family $\overleftarrow{\mathcal{F}} = \{\overleftarrow{\mathcal{F}}_t \mid t \in T\}$ such that $\{\emptyset, \Omega\} \subseteq \overleftarrow{\mathcal{F}}_s \subseteq \overleftarrow{\mathcal{F}}_t \subseteq \Sigma \quad \forall s > t \in T$.

There exist many different filtrations on a probability space $(\Omega, \Sigma, \mathbb{P})$, but not all of them are equally useful for the study of a process X. If we can determine the state X_t from the information given by $\mathcal{F}_t$, we say that X is adapted to the filtration $\mathcal{F}$.

Definition (*Adaptedness*) Given a measurable space $(\Omega, \Sigma, \mathbb{P})$ equipped with a filtration $\mathcal{F}$ and a stochastic process $X : T \times (\Omega, \Sigma, \mathbb{P}) \to (\mathcal{M}, \mathcal{B}(\mathcal{M}))$, X is said to be adapted to the filtration $\mathcal{F}$, if X_t is $\mathcal{F}_t$-measurable for every $t \in T$.

For any process X, we can also define a natural filtration, which is the minimal filtration that contains all the information about the process X.

Definition (*Natural filtration*) Given a stochastic process $X : T \times (\Omega, \Sigma, \mathbb{P}) \to (\mathcal{M}, \mathcal{B}(\mathcal{M}))$, the natural (past) filtration of Σ with respect to X is given by $\vec{\mathcal{F}}^X = \{\vec{\mathcal{F}}_t^X \mid t \in T\}$ with

$$\vec{\mathcal{F}}_t^X = \sigma\big(\{X_s \mid s \le t \in T\}\big) \tag{B.1}$$

and the natural future filtration of Σ with respect to X is given by $\overleftarrow{\mathcal{F}}^X = \{\overleftarrow{\mathcal{F}}_t^X \mid t \in T\}$ with

$$\overleftarrow{\mathcal{F}}_t^X = \sigma\big(\{X_s \mid s \ge t \in T\}\big). \tag{B.2}$$

It immediately follows that any stochastic process X is adapted to its natural filtration.

Definition (*Filtered probability space*) A filtered probability space is a tuple $(\Omega, \Sigma, \mathcal{F}, \mathbb{P})$, where $(\Omega, \Sigma, \mathbb{P})$ is a probability space and $\mathcal{F}$ is a filtration of Σ.

If a process $X : (\Omega, \Sigma, \mathbb{P}) \to (\mathcal{M}, \mathcal{B}(\mathcal{M}))$ is adapted to a filtration $\mathcal{F}$, it can be defined on the filtered probability, such that $X : (\Omega, \Sigma, \mathcal{F}, \mathbb{P}) \to (\mathcal{M}, \mathcal{B}(\mathcal{M}))$.

B.3   Stopping Times

Stopping times are random variables that define the time at which a stochastic process exhibits a certain type of behavior. A typical example is the first time at which the stochastic process starting at $x \in U \subset \mathcal{M}$ hits the boundary ∂U of $U \subset \mathcal{M}$.

Definition (*Stopping time*) Given a set $T \subseteq \mathbb{R}$ and a filtered probability space $(\Omega, \Sigma, \vec{\mathcal{F}}, \mathbb{P})$, a random variable $\tau : (\Omega, \Sigma, \vec{\mathcal{F}}, \mathbb{P}) \to (T, \mathcal{B}(T))$ is a stopping time with respect to the past $\vec{\mathcal{F}}$, if $\{\tau \le t\} \in \vec{\mathcal{F}}_t$ for all $t \in T$.

Definition (*Time reversed stopping time*) Given a set $T \subseteq \mathbb{R}$ and a filtered probability space $(\Omega, \Sigma, \overleftarrow{\mathcal{F}}, \mathbb{P})$, a random variable $\tau : (\Omega, \Sigma, \overleftarrow{\mathcal{F}}, \mathbb{P}) \to (T, \mathcal{B}(T))$ is a stopping time with respect to the future $\overleftarrow{\mathcal{F}}$, if $\{\tau \ge t\} \in \overleftarrow{\mathcal{F}}_t$ for all $t \in T$.

If we stop a process at a stopping time, we obtain a stopped process.

Definition (*Stopped process*) Given a stochastic process $X : \mathcal{T} \times (\Omega, \Sigma, \vec{\mathcal{F}}, \mathbb{P}) \to (\mathcal{M}, \mathcal{B}(\mathcal{M}))$, and a stopping time $\tau : (\Omega, \Sigma, \vec{\mathcal{F}}, \mathbb{P}) \to (\mathcal{T}, \mathcal{B}(\mathcal{T}))$, the process $\{X_t^\tau \mid t \in \mathcal{T}\}$ such that $X_t^\tau = X_{\min\{t, \tau\}}$ is called a stopped process.

Similarly, given a stochastic process $X : \mathcal{T} \times (\Omega, \Sigma, \overleftarrow{\mathcal{F}}, \mathbb{P}) \to (\mathcal{M}, \mathcal{B}(\mathcal{M}))$, and a time reversed stopping time $\tau : (\Omega, \Sigma, \overleftarrow{\mathcal{F}}, \mathbb{P}) \to (\mathcal{T}, \mathcal{B}(\mathcal{T}))$, the process $\{X_t^\tau \mid t \in \mathcal{T}\}$ such that $X_t^\tau = X_{\max\{t, \tau\}}$ is called a stopped process.

B.4 Semi-Martingales

Semi-martingales are processes that can be decomposed into a deterministic drift process and a driftless stochastic process. Semi-martingales are particularly useful, as they form the largest class of stochastic processes for which there exists a stochastic calculus.

Definition (*Càdlàg process*) A process $X : \mathcal{T} \times (\Omega, \Sigma, \mathbb{P}) \to (\mathcal{M}, \mathcal{B}(\mathcal{M}))$ is càdlàg, if

- $\lim_{s \uparrow t} X(s)$ exists a.s. for all $t \in \mathcal{T}$;
- $\lim_{s \downarrow t} X(s) = X(t)$ a.s. for all $t \in \mathcal{T}$.

Definition (*Càglàd process*) A process $X : \mathcal{T} \times (\Omega, \Sigma, \mathbb{P}) \to (\mathcal{M}, \mathcal{B}(\mathcal{M}))$ is càglàd, if

- $\lim_{s \uparrow t} X(s) = X(t)$ a.s. for all $t \in \mathcal{T}$;
- $\lim_{s \downarrow t} X(s)$ exists a.s. for all $t \in \mathcal{T}$.

Definition (*Continuous process*) A process $X : \mathcal{T} \times (\Omega, \Sigma, \mathbb{P}) \to (\mathcal{M}, \mathcal{B}(\mathcal{M}))$ is continuous, if

- $\lim_{s \uparrow t} X(s) = X(t)$ a.s. for all $t \in \mathcal{T}$;
- $\lim_{s \downarrow t} X(s) = X(t)$ a.s. for all $t \in \mathcal{T}$.

Definition (*Finite variation*) A process $X : \mathcal{T} \times (\Omega, \Sigma, \mathbb{P}) \to (\mathbb{R}^n, \mathcal{B}(\mathbb{R}^n))$ has finite variation, if for any finite interval $I \subseteq \mathcal{T}$

$$\mathbb{P}\left[\sup_\Pi \left(\sum_k |X_{t_k}^i - X_{t_{k-1}}^i| \right) < \infty \right] = 1 \quad \forall i \in \{1, ..., n\}, \tag{B.3}$$

where the supremum is taken over all partitions Π of I.

Definition (*Martingale process*) A stochastic process $X : \mathcal{T} \times (\Omega, \Sigma, \vec{\mathcal{F}}, \mathbb{P}) \to (\mathbb{R}^n, \mathcal{B}(\mathbb{R}^n))$ is a martingale with respect to the past $\vec{\mathcal{F}}$, if X is adapted to $\vec{\mathcal{F}}$, X_t is integrable for every $t \in \mathcal{T}$ and

$$\mathbb{E}\left[X_t \,\middle|\, \vec{\mathcal{F}}_s\right] = X_s \quad \forall s < t \in \mathcal{T}. \tag{B.4}$$

Definition (*Time reversed martingale*) A stochastic process $X : \mathcal{T} \times (\Omega, \Sigma, \overleftarrow{\mathcal{F}}, \mathbb{P})$ $\to (\mathbb{R}^n, \mathcal{B}(\mathbb{R}^n))$ is a martingale with respect to the future $\overleftarrow{\mathcal{F}}$, if X is adapted to $\overleftarrow{\mathcal{F}}$, X_t is integrable for every $t \in \mathcal{T}$ and

$$\mathbb{E}\left[X_t \,\middle|\, \overleftarrow{\mathcal{F}}_s\right] = X_s \quad \forall s > t \in \mathcal{T}. \tag{B.5}$$

The martingale property (B.4), which ensure that the process is driftless, is a global property, as it must hold for all pairs $s, t \in \mathcal{T}$, but the property can be localized. All martingales are local martingales, but the opposite is not true.

Definition (*Local martingale*) A stochastic process $X : \mathcal{T} \times (\Omega, \Sigma, \vec{\mathcal{F}}, \mathbb{P}) \to$ $(\mathbb{R}^n, \mathcal{B}(\mathbb{R}^n))$ is a local martingale with respect to the past $\vec{\mathcal{F}}$, if X is adapted to $\vec{\mathcal{F}}$ and there exists a sequence of stopping times $\tau_k : (\Omega, \Sigma, \vec{\mathcal{F}}, \mathbb{P}) \to (\mathcal{T}, \mathcal{B}(\mathcal{T}))$ such that

- $\mathbb{P}(\tau_{k+1} > \tau_k) = 1$;
- $\mathbb{P}[\lim_{k \to \infty} \tau_k = \sup(\mathcal{T})] = 1$;
- the stopped process $X_{\min\{t, \tau_k\}}$ is a martingale with respect to $\vec{\mathcal{F}}$ for every $k \in \mathbb{N}$.

Definition (*Time reversed local martingale*) A stochastic process $X : \mathcal{T} \times (\Omega, \Sigma,$ $\overleftarrow{\mathcal{F}}, \mathbb{P}) \to (\mathbb{R}^n, \mathcal{B}(\mathbb{R}^n))$ is a local martingale with respect to the future $\overleftarrow{\mathcal{F}}$, if X is adapted to $\overleftarrow{\mathcal{F}}$ and there exists a sequence of time reversed stopping times $\tau_k : (\Omega, \Sigma, \overleftarrow{\mathcal{F}}, \mathbb{P}) \to (\mathcal{T}, \mathcal{B}(\mathcal{T}))$ such that

- $\mathbb{P}(\tau_{k+1} < \tau_k) = 1$;
- $\mathbb{P}[\lim_{k \to \infty} \tau_k = \inf(\mathcal{T})] = 1$;
- the stopped process $X_{\max\{t, \tau_k\}}$ is a martingale with respect to $\overleftarrow{\mathcal{F}}$ for every $k \in \mathbb{N}$.

Definition (*Semi-martingale*) A stochastic process $X : \mathcal{T} \times (\Omega, \Sigma, \vec{\mathcal{F}}, \mathbb{P}) \to$ $(\mathbb{R}^n, \mathcal{B}(\mathbb{R}^n))$ is a semi-martingale with respect to the past $\vec{\mathcal{F}}$, if it can be decomposed as

$$X_t = C_t + M_t, \tag{B.6}$$

where C is a càdlàg process of finite variation adapted to $\vec{\mathcal{F}}$ and M is a local martingale with respect to $\vec{\mathcal{F}}$.

Definition (*Time reversed semi-martingale*) A stochastic process $X : \mathcal{T} \times (\Omega, \Sigma,$ $\overleftarrow{\mathcal{F}}, \mathbb{P}) \to (\mathbb{R}^n, \mathcal{B}(\mathbb{R}^n))$ is a semi-martingale with respect to the future $\overleftarrow{\mathcal{F}}$, if it can be decomposed as

$$X_t = C_t + M_t, \tag{B.7}$$

where C is a càglàd process of finite variation adapted to $\overleftarrow{\mathcal{F}}$ and M is a local martingale with respect to $\overleftarrow{\mathcal{F}}$.

Definition (*Two-sided semi-martingale*) A stochastic process $X : \mathcal{T} \times (\Omega, \Sigma, \vec{\mathcal{F}}, \overleftarrow{\mathcal{F}}, \mathbb{P}) \to (\mathbb{R}^n, \mathcal{B}(\mathbb{R}^n))$ is a two-sided semi-martingale with respect to the past $\vec{\mathcal{F}}$ and the future $\overleftarrow{\mathcal{F}}$, if it can be decomposed as

$$X_t = C_{\pm,t} + M_t, \tag{B.8}$$

where C_+ is a càdlàg process of finite variation adapted to $\vec{\mathcal{F}}$, C_- is a càglàd process of finite variation adapted to $\overleftarrow{\mathcal{F}}$ and M is a local martingale with respect to both $\vec{\mathcal{F}}$ and $\overleftarrow{\mathcal{F}}$.

Definition (*Complex valued semi-martingale*) A process $Z : \mathcal{T} \times (\Omega, \Sigma, \mathcal{F}, \mathbb{P}) \to (\mathbb{C}^n, \mathcal{B}(\mathbb{C}^n))$ is a complex semi-martingale with respect to $\mathcal{F}$, if it can be decomposed as $Z = X + \mathrm{i}\, Y$, where $X, Y : \mathcal{T} \times (\Omega, \Sigma, \mathcal{F}, \mathbb{P}) \to (\mathbb{R}^n, \mathcal{B}(\mathbb{R}^n))$ are real semi-martingales with respect to $\mathcal{F}$.

Theorem (Functions of semi-martingales) *Given a semi-martingale* $X : \mathcal{T} \times (\Omega, \Sigma, \mathcal{F}, \mathbb{P}) \to (\mathbb{R}^n, \mathcal{B}(\mathbb{R}^n))$ *and a Borel measurable function* $f \in C^2(\mathbb{R}^n)$, *then* $f \circ X$ *is a semi-martingale.*

Definition (*Manifold valued semi-martingale*) Given a smooth manifold $\mathcal{M}$, a stochastic process $X : \mathcal{T} \times (\Omega, \Sigma, \mathcal{F}, \mathbb{P}) \to (\mathcal{M}, \mathcal{B}(\mathcal{M}))$ is a semi-martingale with respect to $\mathcal{F}$, if $f \circ X$ is a semi-martingale for every $f \in C^\infty(\mathcal{M})$.

B.5 Markov Processes

Markov processes form a class of stochastic processes that obey a memoryless property, which is known as the Markov property.

Definition (*Markov process*) A stochastic process $X : \mathcal{T} \times (\Omega, \Sigma, \vec{\mathcal{F}}, \mathbb{P}) \to (\mathbb{R}^n, \mathcal{B}(\mathbb{R}^n))$ is a Markov process with respect to the past $\vec{\mathcal{F}}$, if X is adapted to $\vec{\mathcal{F}}$, X_t is integrable for every $t \in \mathcal{T}$ and

$$\mathbb{E}[X_t \mid \vec{\mathcal{F}}_s] = \mathbb{E}[X_t \mid X_s] \quad \forall s < t \in \mathcal{T}. \tag{B.9}$$

Definition (*Time reversed Markov process*) A stochastic process $X : \mathcal{T} \times (\Omega, \Sigma, \overleftarrow{\mathcal{F}}, \mathbb{P}) \to (\mathbb{R}^n, \mathcal{B}(\mathbb{R}^n))$ is a Markov process with respect to the future $\overleftarrow{\mathcal{F}}$, if X is adapted to $\overleftarrow{\mathcal{F}}$, X_t is integrable for every $t \in \mathcal{T}$ and

$$\mathbb{E}[X_t \mid \overleftarrow{\mathcal{F}}_s] = \mathbb{E}[X_t \mid X_s] \quad \forall s > t \in \mathcal{T}. \tag{B.10}$$

Definition (*Complex valued Markov process*) A complex stochastic process $Z : \mathcal{T} \times (\Omega, \Sigma, \mathcal{F}, \mathbb{P}) \to (\mathbb{C}^n, \mathcal{B}(\mathbb{C}^n))$ is a Markov process, if it can be decomposed as $Z = X + \mathrm{i}\, Y$, where $X, Y : \mathcal{T} \times (\Omega, \Sigma, \mathcal{F}, \mathbb{P}) \to (\mathcal{M}, \mathcal{B}(\mathcal{M}))$ are real Markov processes.

Definition (*Manifold valued Markov process*) Given a smooth manifold $\mathcal{M}$, a stochastic process $X : T \times (\Omega, \Sigma, \mathcal{F}, \mathbb{P}) \to (\mathcal{M}, \mathcal{B}(\mathcal{M}))$ is a Markov process, if $f \circ X$ is a Markov process for every $f \in C^{\infty}(\mathcal{M})$.

B.6 Quadratic Variation

A characteristic feature of stochastic processes is their non-vanishing quadratic variation. On $\mathbb{R}^n$ this quadratic variation can be defined as a Riemann sum of squares. In the remaining sections of this appendix, we focus on the future directed process, i.e. adapted to the past $\vec{\mathcal{F}}$, but all results can be generalized to a past directed process, i.e. adapted to the future $\overleftarrow{\mathcal{F}}$.

Definition (*Quadratic variation*) Given two stochastic processes $X, Y : T \times (\Omega, \Sigma, \vec{\mathcal{F}}, \mathbb{P}) \to (\mathbb{R}^n, \mathcal{B}(\mathbb{R}^n))$, the quadratic covariation of X and Y is defined by

$$[X, Y]_t = \lim_{||\Pi|| \to 0} \sum_k \left(X_{t_k} - X_{t_{k-1}}\right) \otimes \left(Y_{t_k} - Y_{t_{k-1}}\right), \tag{B.11}$$

where Π is a partition of $[\inf(T), t]$ and $||\Pi||$ is its mesh. Moreover, the processes $[X, X]$ and $[Y, Y]$ are called the quadratic variation of X and Y respectively.

Theorem (Quadratic variation) *Given two semi-martingales* $X, Y : T \times (\Omega, \Sigma, \vec{\mathcal{F}}, \mathbb{P}) \to (\mathbb{R}^n, \mathcal{B}(\mathbb{R}^n))$, *the quadratic covariation* $[X, Y]$ *is a* $(\mathbb{R}^{n \times n})$-*valued càdlàg process of finite variation adapted to* $\vec{\mathcal{F}}$.

Theorem (Properties of quadratic variation I) *Given semi-martingales* $X, Y, Z : T \times (\Omega, \Sigma, \vec{\mathcal{F}}, \mathbb{P}) \to (\mathbb{R}^n, \mathcal{B}(\mathbb{R}^n))$, *the quadratic covariation satisfies the following properties*

- *(symmetry):* $[X, Y] = [Y, X]$;
- *(bilinearity):* $[a X + b Y, Z] = a [X, Z] + b [Y, Z]$ *for all* $a, b \in \mathbb{R}$;
- *(positivity):* $[X, X]_t - [X, X]_s$ *is a.s. positive semi-definite for all* $s \leq t \in T$.

A stochastic processes $X : T \times (\Omega, \Sigma, \mathcal{F}, \mathbb{P}) \to (\mathbb{R}^n, \mathcal{B}(\mathbb{R}^n))$ is not necessarily continuous, but, if it is a semi-martingale, the number of discontinuities is countable, such that

$$X_t = X_t^c + \sum_k \Delta X_{t_k}, \tag{B.12}$$

where X^c denotes the continuous part of X, ΔX_{t_k} the size of the jumps and k ranges over all jumps up to time t. Similarly, for the quadratic variation of two semi-martingales X, Y, we can write

$$[X, Y]_t = [X, Y]_t^c + \sum_k \Delta[X, Y]_{t_k}, \tag{B.13}$$

where

$$\Delta[X, Y]_{t_k} = \Delta X_{t_k} \Delta Y_{t_k}. \tag{B.14}$$

Using this decomposition, we can formulate a few more properties.

Lemma (Properties quadratic variation III) *Given semi-martingales* $X, Y : \mathcal{T} \times (\Omega, \Sigma, \mathcal{F}, \mathbb{P}) \to (\mathbb{R}^n, \mathcal{B}(\mathbb{R}^n))$, *the quadratic covariation satisfies the following properties*

- $[X, Y]_t^c = 0$, *if* X_t *or* Y_t *has finite variation;*
- $\Delta[X, Y]_{t_k} = 0$, *if* $\Delta X_{t_k} = 0$ *or* $\Delta Y_{t_k} = 0$.

Corollary (Properties quadratic variation IV) *Given semi-martingales* $X, Y, Z : \mathcal{T} \times (\Omega, \Sigma, \mathcal{F}, \mathbb{P}) \to (\mathbb{R}^n, \mathcal{B}(\mathbb{R}^n))$, *the quadratic covariation satisfies the following properties*

- $[X, Y]$ *is continuous, if at least one of the processes* X, Y *is continuous;*
- $[[X, Y], Z] = 0$, *if at least one of the processes* X, Y, Z *is continuous.*

In Appendix D, we show that the sum (B.11) used in the definition of the quadratic variation is intrinsic on a pseudo-Riemannian manifold. This allows to construct a manifold valued quadratic variation process $[X, Y] : \mathcal{T} \times (\Omega, \Sigma, \mathcal{F}, \mathbb{P}) \to (\mathcal{M}^{\otimes 2}, \mathcal{B}(\mathcal{M})^{\otimes 2})$ on all pseudo-Riemannian manifolds $(\mathcal{M}, g)$, which satisfies the same properties.

Furthermore, since all complex valued processes can be decomposed into real valued processes, one can straightforwardly generalize the quadratic variation and its properties to complex valued processes.

We emphasize that all properties of the quadratic variation rely on the fact that the semi-martingales are adapted to the same filtration $\vec{F}$. If X and Y cannot be adapted to a common filtration, the limit (B.11) may not exist. This fact is crucial for the discussion in Sect. 3.6.

B.7 Lévy Processes

Lévy processes form a class of Markov processes, and can be regarded as the continuous time analog of the random walk.

Definition (*Lévy process*) A stochastic process $X : \mathcal{T} \times (\Omega, \Sigma, \mathbb{P}) \to (\mathbb{R}^n, \mathcal{B}(\mathbb{R}^n))$ is a Lévy process, if it is continuous in probability and has independent stationary increments, i.e.

- $\forall \epsilon > 0, t \in \mathcal{T}, \lim_{s \to t} \mathbb{P}(\|X_s - X_t\| \geq \epsilon) = 0$;
- $\forall t_1 < t_2 < t_3 < t_4 \in \mathcal{T}, (X_{t_4} - X_{t_3}) \perp\!\!\!\perp (X_{t_2} - X_{t_1})$;

- $\forall t_1, t_2, t_3, t_4 \in \mathcal{T}$, s.t. $|t_4 - t_3| = |t_2 - t_1|$, $d\mu_{(X_{t_4} - X_{t_3})} = d\mu_{(X_{t_2} - X_{t_1})}$.

Definition (*Complex Lévy process*) A stochastic process $Z : \mathcal{T} \times (\Omega, \Sigma, \mathbb{P}) \rightarrow (\mathbb{C}^n, \mathcal{B}(\mathbb{C}^n))$ is a complex valued Lévy process, if it can be decomposed as $Z = X + iY$, where $X, Y : \mathcal{T} \times (\Omega, \Sigma, \mathbb{P}) \rightarrow (\mathbb{R}^n, \mathcal{B}(\mathbb{R}^n))$ are independent real valued Lévy processes.

Definition (*Lévy process w.r.t. a filtration*) Given a Lévy process $X : \mathcal{T} \times (\Omega, \Sigma, \mathbb{P}) \rightarrow (\mathbb{R}^n, \mathcal{B}(\mathbb{R}^n))$, we call $X : \mathcal{T} \times (\Omega, \Sigma, \mathcal{F}, \mathbb{P}) \rightarrow (\mathbb{R}^n, \mathcal{B}(\mathbb{R}^n))$ a Lévy process with respect to the filtration $\mathcal{F}$, if X is adapted to $\mathcal{F}$ and $(X_t - X_s) \perp\!\!\!\perp \mathcal{F}_s$ for all $t > s \in \mathcal{T}$ ($s < t$, if $\mathcal{F}$ is a time reversed filtration).

Theorem (Lévy process is Markov) *A Lévy process* $X : \mathcal{T} \times (\Omega, \Sigma, \mathcal{F}, \mathbb{P}) \rightarrow (\mathbb{R}^n, \mathcal{B}(\mathbb{R}^n))$ *is a Markov process w.r.t.* $\mathcal{F}$.

Theorem (Lévy process is semi-martingale) *A Lévy process* $X : \mathcal{T} \times (\Omega, \Sigma, \mathcal{F}, \mathbb{P}) \rightarrow (\mathbb{R}^n, \mathcal{B}(\mathbb{R}^n))$ *is a semi-martingale w.r.t.* $\mathcal{F}$.

B.8 Wiener Processes

The Wiener process is a continuous Lévy process that is best known for its application in describing Brownian motion. For this reason, the terms Wiener process and Brownian motion are often used interchangeably.

Definition (*Wiener process*) A stochastic process $X : \mathcal{T} \times (\Omega, \Sigma, \mathbb{P}) \rightarrow (\mathbb{R}^n, \mathcal{B}(\mathbb{R}^n))$ is a Wiener process (a.k.a. Brownian motion), if it is almost surely continuous and has independent normally distributed increments, i.e.

- $\mathbb{P}\left(\lim_{s \to t} \|X_s - X_t\| = 0\right) = 1$;
- $\forall t_1 < t_2 < t_3 < t_4 \in \mathcal{T}$, $(X_{t_4} - X_{t_3}) \perp\!\!\!\perp (X_{t_2} - X_{t_1})$;
- $\forall t_1, t_2 \in \mathcal{T}$, $(X_{t_2} - X_{t_1}) \sim \mathcal{N}(0, \alpha \, \delta^{ij} \, |t_2 - t_1|)$,

where $\mathcal{N}(\mu, \sigma^2)$ denotes the normal distribution with mean $\mu \in \mathbb{R}^n$ and covariance matrix $\sigma^2 \in \mathbb{R}^{n \times n}$. Furthermore, $\alpha \in (0, \infty)$ is a scaling parameter.

Definition (*Complex Wiener process*) A stochastic process $Z : \mathcal{T} \times (\Omega, \Sigma, \mathbb{P}) \rightarrow (\mathbb{C}^n, \mathcal{B}(\mathbb{C}^n))$ is a complex valued Wiener process, if it can be decomposed as $Z = X + iY$, where $X, Y : \mathcal{T} \times (\Omega, \Sigma, \mathbb{P}) \rightarrow (\mathbb{R}^n, \mathcal{B}(\mathbb{R}^n))$ are independent real valued Wiener processes.

Theorem (Wiener process is Lévy) *A Wiener process* $X : \mathcal{T} \times (\Omega, \Sigma, \mathbb{P}) \rightarrow (\mathbb{R}^n, \mathcal{B}(\mathbb{R}^n))$ *is a continuous Lévy process.*

Theorem (Wiener process is martingale) *A Wiener process* $X : \mathcal{T} \times (\Omega, \Sigma, \mathcal{F}, \mathbb{P}) \rightarrow (\mathbb{R}^n, \mathcal{B}(\mathbb{R}^n))$ *is a martingale with respect to* $\mathcal{F}$.

There exists an alternative definition of the Wiener process, known as the Lévy characterization [11]. In this equivalent definition, the requirement that the increments are normally distributed is replaced by a condition on the quadratic variation.

Theorem (Lévy characterization of Brownian motion) *Given a continuous local martingale* $X : T \times (\Omega, \Sigma, \vec{\mathcal{F}}, \mathbb{P}) \to (\mathbb{R}^n, \mathcal{B}(\mathbb{R}^n))$ *and a continuous finite variation process* $\sigma^2 : T \to \mathbb{R}^{n \times n}$, *such that* $\sigma_t^2 - \sigma_s^2$ *is positive semi-definite for all* $t \geq s \in T$, *then the following statements are equivalent*

- *(i)* $(X_t - X_s) \perp\!\!\!\perp \vec{\mathcal{F}}_s$ *and* $(X_t - X_s) \sim \mathcal{N}(0, \sigma_t^2 - \sigma_s^2)$ *for all* $t \geq s \in T$;
- *(ii)* $X_t \otimes X_t - \sigma_t^2$ *is a local martingale;*
- *(iii)* X *has a quadratic variation given by* $[X, X]_t = \sigma_t^2$.

In particular, if $(\sigma_t^2 - \sigma_s^2)^{ij} = \alpha \, \delta^{ij} |t - s|$, *(i) implies that* X *is a Brownian motion (Wiener process) with respect to* $\vec{\mathcal{F}}$ *with scaling parameter* α.

B.9 L^2-spaces

The definitions of the L^2-norm and the L^2-spaces encountered in Appendix A.7 can be generalized to the context of stochastic processes.

Definition (L^2-*norm*) Given a stochastic process $X : T \times (\Omega, \Sigma, \mathbb{P}) \to (\mathbb{K}^n, \mathcal{B}(\mathbb{K}^n))$ with $\mathbb{K} \in \{\mathbb{R}, \mathbb{C}\}$, the L^2-norm for X is defined by

$$||X|| = \sqrt{\mathbb{E}\left[\int_T \delta_{ij} \overline{X}_t^i X_t^j \, dt\right]}. \tag{B.15}$$

Definition (L^2-*space*) Given a probability space $(\Omega, \Sigma, \mathbb{P})$ and a set $T \subseteq \mathbb{R}$, the spaces

$$L_T^2(\Omega, \Sigma, \mathbb{P}) = \left\{X : T \times (\Omega, \Sigma, \mathbb{P}) \to (\mathbb{K}^n, \mathcal{B}(\mathbb{K}^n)) \,\Big|\, ||X|| < \infty\right\}$$

with $\mathbb{K} \in \{\mathbb{R}, \mathbb{C}\}$ are the real and complex L^2-space over $T \times (\Omega, \Sigma, \mathbb{P})$.

Definition (*Velocity*) Given a stochastic process $X : T \times (\Omega, \Sigma, \mathbb{P}) \to (\mathcal{M}, \mathcal{B}(\mathcal{M}))$, we define the forward Itô velocity by

$$v_+(X_t, t) = \lim_{dt \to 0} \mathbb{E}\left[\frac{X_{t+dt} - X_t}{dt} \,\Big|\, X_t\right], \tag{B.16}$$

the backward Itô velocity by

$$v_-(X_t, t) = \lim_{dt \to 0} \mathbb{E}\left[\frac{X_t - X_{t-dt}}{dt} \,\Big|\, X_t\right], \tag{B.17}$$

the Stratonovich velocity by

$$v_\circ(X_t, t) = \frac{1}{2}\Big[v_+(X_t, t) + v_-(X_t, t)\Big],$$ (B.18)

and the second order velocity by

$$v_2(X_t, t) = \lim_{dt \to 0} \mathbb{E}\left[\frac{(X_{t+dt} - X_t)(X_{t+dt} - X_t)}{dt}\,\Big|\, X_t\right].$$ (B.19)

Definition *(Length)* Given a Riemannian manifold $(\mathcal{M}, g)$ and a stochastic process $X : \mathcal{T} \times (\Omega, \Sigma, \mathbb{P}) \to (\mathcal{M}, \mathcal{B}(\mathcal{M}))$, we define the length of X by

$$\Delta X_T = \mathbb{E}\left[\int_\mathcal{T} \sqrt{g(v_\circ, v_\circ)(X_t, t)}\, dt\right].$$ (B.20)

Definition *(Time-like, null-like and space-like)* Given a Lorentzian manifold $(\mathcal{M}, g)$ with $(-+...+)$ signature and a stochastic process $X : \mathcal{T} \times (\Omega, \Sigma, \mathbb{P}) \to (\mathcal{M}, \mathcal{B}(\mathcal{M}))$, we call the process X

- time-like, if $\mathbb{E}\left[g(v_\circ, v_\circ)(X_t, t)\right] < 0 \ \ \forall t \in \mathcal{T}$,
- null-like, if $\mathbb{E}\left[g(v_\circ, v_\circ)(X_t, t)\right] = 0 \ \ \forall t \in \mathcal{T}$,
- space-like, if $\mathbb{E}\left[g(v_\circ, v_\circ)(X_t, t)\right] > 0 \ \ \forall t \in \mathcal{T}$.

Definition *(Proper time)* Given a Lorentzian manifold $(\mathcal{M}, g)$ with $(-+...+)$ signature and a time-like stochastic process $X : \mathcal{T} \times (\Omega, \Sigma, \mathbb{P}) \to (\mathcal{M}, \mathcal{B}(\mathcal{M}))$, we define the proper time of X by

$$\Delta\tau(X_T) = \mathbb{E}\left[\int_\mathcal{T} \sqrt{-g(v_\circ, v_\circ)(X_t, t)}\, dt\right].$$ (B.21)

Definition *(Proper distance)* Given a Lorentzian manifold $(\mathcal{M}, g)$ with $(-+...+)$ signature and a space-like stochastic process $X : \mathcal{T} \times (\Omega, \Sigma, \mathbb{P}) \to (\mathcal{M}, \mathcal{B}(\mathcal{M}))$, we define the proper length of X by

$$\Delta s(X_T) = \mathbb{E}\left[\int_I \sqrt{g(v_\circ, v_\circ)(X_t, t)}\, dt\right].$$ (B.22)

Definition *(Induced L^2-norm)* Given a stochastic process $X : \mathcal{T} \times (\Omega, \Sigma, \mathbb{P}) \to (\mathcal{M}, \mathcal{B}(\mathcal{M}))$ and a Borel measurable function $f : \mathcal{M} \times \mathcal{T} \to \mathbb{K}$ with $\mathbb{K} \in \{\mathbb{R}, \mathbb{C}\}$, the L^2-norm for $f \circ X$ is defined by

$$\|f(X)\| = \sqrt{\int_\mathcal{T} \mathbb{E}\Big[|f(X_t, t)|^2\Big]\, dt}\,.$$ (B.23)

Definition (*Induced L^2-space*) Given a stochastic process $X : T \times (\Omega, \Sigma, \mathbb{P}) \to (\mathcal{M}, \mathcal{B}(\mathcal{M}))$, the spaces

$$L_T^2(\mathcal{M}, \mathcal{B}(\mathcal{M}), \mu_X) = \left\{ f : (\mathcal{M}, \mathcal{B}(\mathcal{M})) \times T \to (\mathbb{K}, \mathcal{B}(\mathbb{K})) \,\middle|\, \|f(X)\| < \infty \right\}$$
(B.24)

with $\mathbb{K} \in \{\mathbb{R}, \mathbb{C}\}$ are the real and complex L^2-space over $(\mathcal{M}, \mathcal{B}(\mathcal{M}), \mu_X) \times T$.

Theorem (Properties of L^2-spaces) *L^2-spaces have the following properties:*

- *L^2 is a Hilbert space with inner product*
  - *$X \cdot Y = \mathbb{E}[\int_T \delta_{ij} \overline{X}^i Y^j \, dt]$ on $L_T^2(\Omega, \Sigma, \mathbb{P})$,*
  - *$\langle f, g \rangle_{\mu_X} = \int_T \int_{\mathcal{M}} \overline{f(x, t)} \, g(x, t) \, d\mu_{X_t}(x, t) \, dt$ on $L_T^2(\mathcal{M}, \mathcal{B}(\mathcal{M}), \mu_X)$;*
- *L^2 is self-dual, i.e. $L^2 \cong (L^2)^*$.*

B.10 Generating Functionals

The definitions of the characteristic function and moment generating function encountered in Appendix A.7 can be generalized to the context of stochastic processes.

Definition (*Moment generating functional*) Given a real stochastic process $X : T \times (\Omega, \Sigma, \mathbb{P}) \to (\mathbb{R}^n, \mathcal{B}(\mathbb{R}^n))$ and a family of linear forms $a = \{a_t \in [L^2(\Omega, \Sigma, \mathbb{P})]^* \mid t \in T\}$, the moment generating functional of X is defined by

$$M_X(a) := \mathbb{E}\left[e^{\int_T a_t(X_t) \, dt} \right],$$
(B.25)

where $\|a_t\| < \rho_t$ with ρ_t the radius of convergence of M_{X_t}.

Given a complex stochastic process $Z : T \times (\Omega, \Sigma, \mathbb{P}) \to (\mathbb{C}^n, \mathcal{B}(\mathbb{C}^n))$ and a family of linear forms $a = \{a_t \in [L^2(\Omega, \Sigma, \mathbb{P})]^* \mid t \in T\}$, the moment generating functional of Z is defined by

$$M_Z(a) := \mathbb{E}\left[e^{\int_T \overline{a}_t(Z_t) \, dt} \right],$$
(B.26)

where $\|a_t\| < \rho_t$ with ρ_t the radius of convergence of M_{Z_t}.

Definition (*Characteristic functional*) Given a real stochastic process $X : T \times (\Omega, \Sigma, \mathbb{P}) \to (\mathbb{R}^n, \mathcal{B}(\mathbb{R}^n))$ and a family of linear forms $a = \{a_t \in [L^2(\Omega, \Sigma, \mathbb{P})]^* \mid t \in T\}$, the characteristic functional of X is defined by

$$\varphi_X(a) := \mathbb{E}\left[e^{i \int_T a_t(X_t) \, dt} \right].$$
(B.27)

Given a complex stochastic process $Z : \mathcal{T} \times (\Omega, \Sigma, \mathbb{P}) \to (\mathbb{C}^n, \mathcal{B}(\mathbb{C}^n))$ and a family of linear forms $a = \{a_t \in [L^2(\Omega, \Sigma, \mathbb{P})]^* \mid t \in \mathcal{T}\}$, the characteristic functional of Z is defined by

$$\varphi_Z(a) := \mathbb{E}\left[e^{i \int_{\mathcal{T}} \mathrm{Re}[\bar{a}_t(Z_t)]\, dt} \right]. \tag{B.28}$$

Appendix C
Review of Stochastic Calculus

Ordinary analysis and its associated calculus allow to describe deterministic trajectories of finite variation, but it is no longer applicable when the trajectories are stochastic. In this case, one must resort to stochastic analysis and stochastic calculus.

A main difficulty in stochastic analysis is the fact that generic stochastic processes are almost surely not differentiable. As a consequence, one cannot unambiguously define a derivative of a stochastic process, which prevents the construction of a differential calculus. One can, however, construct an integral along any stochastic trajectory that is a semi-martingale, which induces an integral calculus. In this appendix, we review some elementary properties of this calculus.

Given a semi-martingale $X : T \times (\Omega, \Sigma, \mathcal{F}, \mathbb{P}) \to (\mathbb{R}, \mathcal{B}(\mathbb{R}))$ and a Borel measurable function $f : \mathbb{R} \times T \to \mathbb{R}$ that is twice continuously differentiable, one can construct an integral along X in various ways. The first is called the *Stratonovich integral* and is given by

$$\int_T f(X_t, t) \circ dX_t := \lim_{||\Pi|| \to 0} \sum_k \frac{1}{2} \big[f(X_{t_k}, t_k) + f(X_{t_{k+1}}, t_{k+1}) \big] \big[X_{t_{k+1}} - X_{t_k} \big],$$

(C.1)

where Π is a partition of T and $||\Pi||$ is its mesh. The second is called the *Itô integral* and is given by

$$\int_T f(X_t, t) \, d_+ X_t := \lim_{||\Pi|| \to 0} \sum_k f(X_{t_k}, t_k) \big[X_{t_{k+1}} - X_{t_k} \big].$$

(C.2)

A third is called a *backward Itô integral*, and is given by

$$\int_T f(X_t, t) \, d_- X_t := \lim_{||\Pi|| \to 0} \sum_k f(X_{t_{k+1}}, t_{k+1}) \big[X_{t_{k+1}} - X_{t_k} \big].$$

(C.3)

F. Kuipers, *Stochastic Mechanics*, SpringerBriefs in Physics, https://doi.org/10.1007/978-3-031-31448-3

Finally, in contrast to deterministic processes, stochastic process have a non-vanishing quadratic variation.[1] This allows to define a fourth integral, which is the *integral over the quadratic variation* given by

$$\int_T f(X_t, t)\, d[X, X]_t := \lim_{||\Pi|| \to 0} \sum_k f(X_{t_k}, t_k) \left[X_{t_{k+1}} - X_{t_k} \right]^2. \qquad (C.4)$$

In the remainder of this appendix and throughout the book, we will assume that the semi-martingale X is continuous.[2] Starting from their respective definitions, one can derive a relation between the various integrals that is given by

$$\int_T f(X_t, t) \circ dX_t = \int_T f(X_t, t)\, d_+ X_t + \frac{1}{2} \int_T \frac{\partial}{\partial x} f(X_t, t)\, d[X, X]_t$$

$$= \int_T f(X_t, t)\, d_- X_t - \frac{1}{2} \int_T \frac{\partial}{\partial x} f(X_t, t)\, d[X, X]_t. \qquad (C.5)$$

It is common in stochastic calculus to work in a differential notation, where the integral sign is dropped. In this notation, these relations are expressed as

$$f(X_t, t) \circ dX_t = f(X_t, t)\, d_+ X_t + \frac{1}{2} \partial_x f(X_t, t)\, d[X, X]_t$$

$$= f(X_t, t)\, d_- X_t - \frac{1}{2} \partial_x f(X_t, t)\, d[X, X]_t. \qquad (C.6)$$

In the remainder of this appendix and throughout the book, we will make use this differential notation. In addition, we denote Stratonovich integrals by $d_\circ$ instead of $\circ d$.

We will now discuss some properties of the various integrals that can be derived from their respective definitions. The Stratonovich integral satisfies the ordinary chain rule and Leibniz rule, i.e. for any function f, g with the same properties as before we have

$$d_\circ f = \partial_t f\, dt + \partial_x f\, d_\circ X, \qquad (C.7)$$

$$d_\circ(f\, g) = f\, d_\circ g + g\, d_\circ f. \qquad (C.8)$$

The Itô integral, on the other hand, satisfies Itô's lemma and a modified Leibniz rule given by

---

[1] More precisely, in a deterministic theory the quadratic variation is non-vanishing for discontinuous processes only, while it is non-vanishing for all processes in a stochastic theory.

[2] All results can be generalized to processes that are càdlàg or càglàd, but they obtain corrections due to the presence of jump discontinuities.

$$d_+ f = \partial_t f \, dt + \partial_x f \, d_+ X + \frac{1}{2} \partial_x^2 f \, d[X, X], \qquad (C.9)$$

$$d_+ (f \, g) = f \, d_+ g + g \, d_+ f + d[f, g]. \qquad (C.10)$$

Similarly, for the backward Itô integral, one obtains

$$d_- f = \partial_t f \, dt + \partial_x f \, d_- X - \frac{1}{2} \partial_x^2 f \, d[X, X], \qquad (C.11)$$

$$d_- (f \, g) = f \, d_- g + g \, d_- f - d[f, g]. \qquad (C.12)$$

Furthermore, the quadratic variation satisfies a symmetry property together with a chain and product rule. These are given by

$$d[f, g] = d[g, f], \qquad (C.13)$$

$$d[f, g] = \partial_x f \, \partial_x g \, d[X, X], \qquad (C.14)$$

$$d[f \, g, h] = f \, d[g, h] + g \, d[f, h]. \qquad (C.15)$$

Itô integrals along local martingales are themselves local martingales. Itô integrals thus satisfy the following martingale property:

$$\mathbb{E}\left[\int_s^t f(X_r, r) \, d_+ M_r \,\Big|\, \vec{\mathcal{F}}_s \right] = 0 \qquad \forall \, s < t \in \mathcal{T},$$

$$\mathbb{E}\left[\int_s^t f(X_r, r) \, d_- M_r \,\Big|\, \overleftarrow{\mathcal{F}}_t \right] = 0 \qquad \forall \, s < t \in \mathcal{T}, \qquad (C.16)$$

where it is assumed that M is a local martingale with respect to the past $\vec{\mathcal{F}}$ in the first line and with respect to the future $\overleftarrow{\mathcal{F}}$ in the second line and that X is adapted to these filtrations.

All results from this appendix can be extended to higher dimensional semi-martingales. For a n-dimensional real continuous semi-martingale $X : \mathcal{T} \times (\Omega, \Sigma, \mathcal{F}, \mathbb{P}) \to (\mathbb{R}^n, \mathcal{B}(\mathbb{R}^n))$ the Stratonovich and Itô integrals can be defined for every integrable and Borel measurable form $f \in T^* \mathbb{R}^d$ by

$$\int_{\mathcal{T}} f(d_\circ X_t) = \int_{\mathcal{T}} f_i(X_t) \, d_\circ X_t^i \,,$$

$$\int_{\mathcal{T}} f(d_\pm X_t) = \int_{\mathcal{T}} f_i(X_t) \, d_\pm X_t^i \,. \qquad (C.17)$$

Similarly, the integral over the quadratic variation can be defined using bilinear forms $g \in T^2(T^* \mathbb{R}^d)$ such that

$$\int_T g(d_\circ X_t, d_\circ X_t) = \int_T g_{ij}(X_t)\, d[X^i, X^j]_t ,$$

$$\int_T g(d_\pm X_t, d_\pm X_t) = \int_T g_{ij}(X_t)\, d[X^i, X^j]_t . \tag{C.18}$$

All the results can also be extended to complex valued continuous stochastic processes Z, using their decomposition $Z = X + i\,Y$ into real stochastic processes. For the Stratonovich and Itô integrals one then obtains

$$\int_T f(Z_t)\, d_\circ Z_t = \int_T f(Z_t)\, d_\circ X_t + i \int_T f(Z_t)\, d_\circ Y_t \tag{C.19}$$

$$\int_T f(Z_t)\, d_\pm Z_t = \int_T f(Z_t)\, d_\pm X_t + i \int_T f(Z_t)\, d_\pm Y_t \tag{C.20}$$

and for the quadratic variation one obtains

$$\int_T f(Z_t)\, d[Z, Z]_t = \int_T f(Z_t)\, d[X, X]_t - \int_T f(Z_t)\, d[Y, Y]_t + 2i \int_T f(Z_t)\, d[X, Y]_t , \tag{C.21}$$

where we used the symmetry of the quadratic variation.

Appendix D
Second Order Geometry

Second order geometry is a geometrical framework that allows to extend stochastic calculus to manifolds. Stratonovich calculus is more adaptable for such an extension than Itô calculus, since the Stratonovich formalism obeys the ordinary Leibniz' rule and chain rule: for any $f, g \in C^\infty(\mathcal{M})$ and $h \in C^2(\mathbb{R})$

$$d_\circ(f\, g) = f\, d_\circ g + g\, d_\circ f \,,$$
$$d_\circ(h \circ f) = \left(h' \circ f\right) d_\circ f \,.$$

In Itô calculus, on the other hand, Leibniz rule is violated and the chain rule is replaced by Itô's lemma:

$$d_\pm(f\, g) = f\, d_\pm g + g\, d_\pm f \pm d[f, g]\,,$$
$$d(h \circ f) = \left(h' \circ f\right) d_\pm f \pm \frac{1}{2} \left(h'' \circ f\right) d[f, f] \,.$$

Nevertheless, since Itô calculus satisfies the martingale property (C.16), it is worth extending the framework to manifolds. If the manifold is twice continuously differentiable and equipped with an affine connection, this can be done using second order geometry, which was developed by Schwartz and Meyer [68–71].

The idea of second order geometry is to incorporate the violation of Leibniz' rule, which is a property of functions defined on the manifold, into the underlying geometry. In this appendix, we review some aspects of this framework for smooth manifolds.

We will consider a smooth manifold equipped with affine connection $(\mathcal{M}, \Gamma)$ and smooth functions $f \in C^\infty(\mathcal{M})$. In classical geometry, the differential of such a function along a trajectory $X : \mathcal{T} \to \mathcal{M}$ is given by

© The Editor(s) (if applicable) and The Author(s), under exclusive license to Springer Nature Switzerland AG 2023
F. Kuipers, *Stochastic Mechanics*, SpringerBriefs in Physics,
https://doi.org/10.1007/978-3-031-31448-3

$$df(X_t, t) = \partial_t f \, dt + \partial_i f \, dX_t^i + \mathcal{O}(dt^2)$$
$$= \left[\partial_t f + v^i \, \partial_i f \right] dt + \mathcal{O}(dt^2) , \tag{D.1}$$

where $v(X_t, t)$ is a velocity field defined on the tangent bundle $T\mathcal{M}$.

In a stochastic theory, the trajectory X becomes a stochastic process and a similar expression can be obtained for semi-martingales in the Itô formulation:

$$\mathbb{E}\left[d_\pm f(X_t, t) \,\middle|\, X_t \right] = \mathbb{E}\left[\partial_t f \, dt + \partial_i f \, d_\pm X_t^i \pm \frac{1}{2} \partial_j \partial_i f \, d[X^i, X^j]_t + o(dt) \,\middle|\, X_t \right]$$
$$= \left[\partial_t f + v_\pm^i \, \partial_i f \pm \frac{1}{2} v_2^{ij} \, \partial_j \partial_i f \right] dt + o(dt) , \tag{D.2}$$

where, cf. Appendix B.9,

$$v_\pm(X_t, t) = \lim_{dt \to 0} \mathbb{E}\left[\frac{d_\pm X_t}{dt} \,\middle|\, X_t \right],$$
$$v_2(X_t, t) = \lim_{dt \to 0} \mathbb{E}\left[\frac{d[X, X]_t}{dt} \,\middle|\, X_t \right]. \tag{D.3}$$

In a deterministic theory, terms appearing at $\mathcal{O}(dt)$ are associated with velocities, terms appearing at $\mathcal{O}(dt^2)$ with accelerations etc. Second order geometry preserves this idea and thus interprets both the fields $v_\pm$ and v_2 as a velocity. Here, $v_\pm$ is a velocity field in the usual sense, while v_2 is a velocity field associated with the quadratic variation of X. Thus, by the Lévy characterization, cf. Appendix B.8, if X is a Wiener process, $v_\pm$ determines a velocity associated to the expectation value $\mathbb{E}[X_t]$, while v_2 determines the velocity associated to the variance $\mathrm{Var}(X_t) = \mathbb{E}[X_t^2] - \mathbb{E}[X_t]^2$.

As a consequence, on a n-dimensional manifold, second order velocity fields (v^i, v^{jk}) have $\frac{n(n+3)}{2}$ degrees of freedom, where n degrees of freedom are encoded in v^i and the remaining $\frac{n(n+1)}{2}$ degrees of freedom in the symmetric object v^{jk}. It follows that, at every point $x \in \mathcal{M}$, the n-dimensional first order tangent space $T_x\mathcal{M}$ must be extended to a $\frac{n(n+3)}{2}$-dimensional second order tangent space $T_{2,x}\mathcal{M}$. Then, in a local coordinate chart, second order vectors $v \in T_{2,x}\mathcal{M}$ can be represented with respect to their canonical basis as[3]

$$v = v^i \, \partial_i + \frac{1}{2} v^{jk} \, \partial_{jk} , \tag{D.4}$$

whereas first order vectors $v \in T_x\mathcal{M}$ are given by

$$v = v^i \, \partial_i . \tag{D.5}$$

---

[3] The factor $\frac{1}{2}$ is not universal in the definition of second order vectors. Some works, e.g. Refs. [41, 71], include it, while other works, e.g. Refs. [66, 70], do not.

For the same reason, for any $x \in \mathcal{M}$, the first order cotangent space $T_x^*\mathcal{M}$ must be extended to a second order cotangent space $T_{2,x}^*\mathcal{M}$. In a local coordinate chart, first order forms $\omega \in T_x^*\mathcal{M}$ are given with respect to their canonical basis by

$$\omega = \omega_i \, dx^i \tag{D.6}$$

and second order forms $\omega \in T_{2,x}^*\mathcal{M}$ are given by

$$\omega = \omega_i \, d_2 x^i + \frac{1}{2} \omega_{ij} \, d[x^i, x^j]. \tag{D.7}$$

The duality pairing of first order vectors $v \in T_x\mathcal{M}$ with forms $\omega \in T_x^*\mathcal{M}$ is given by

$$\langle \omega, v \rangle = \omega_i \, v^i. \tag{D.8}$$

Similarly, the duality pairing for second order vectors $v \in T_{2,x}\mathcal{M}$ with second order forms $\omega \in T_{2,x}^*\mathcal{M}$ is

$$\langle \omega, v \rangle = \omega_i \, v^i + \frac{1}{2} \omega_{ij} \, v^{ij}, \tag{D.9}$$

and can be derived from the duality pairing of the basis elements

$$\begin{aligned}
\langle d_2 x^i, \partial_j \rangle &= \partial_j x^i &&= \delta_j^i, \\
\langle d_2 x^i, \partial_{jk} \rangle &= \partial_j \partial_k x^i &&= 0, \\
\langle d[x^i, x^j], \partial_k \rangle &= \partial_k(x^i x^j) - x^i \partial_k x^j - x^j \partial_k x^i &&= 0, \\
\langle d[x^i, x^j], \partial_{kl} \rangle &= \partial_k \partial_l(x^i x^j) - x^i \partial_k \partial_l x^j - x^j \partial_k \partial_l x^i &&= \delta_k^i \delta_l^j + \delta_l^i \delta_k^j.
\end{aligned} \tag{D.10}$$

Second order vectors and forms do not transform in a covariant manner. Indeed, one can easily verify that the active coordinate transformation laws for vectors are

$$\begin{aligned}
v^i \to \tilde{v}^i &= v^k \frac{\partial \tilde{x}^i}{\partial x^k} + \frac{1}{2} v^{kl} \frac{\partial^2 \tilde{x}^i}{\partial x^k \partial x^l}, \\
v^{ij} \to \tilde{v}^{ij} &= v^{kl} \frac{\partial \tilde{x}^i}{\partial x^k} \frac{\partial \tilde{x}^j}{\partial x^l},
\end{aligned} \tag{D.11}$$

and for forms they are given by

$$\begin{aligned}
\omega_i \to \tilde{\omega}_i &= \omega_k \frac{\partial x^k}{\partial \tilde{x}^i}, \\
\omega_{ij} \to \tilde{\omega}_{ij} &= \omega_k \frac{\partial^2 x^k}{\partial \tilde{x}^i \partial \tilde{x}^j} + \omega_{kl} \frac{\partial x^k}{\partial \tilde{x}^i} \frac{\partial x^l}{\partial \tilde{x}^j}.
\end{aligned} \tag{D.12}$$

Similarly, the passive transformation laws are given by

$$\begin{aligned}
\partial_i \to \tilde{\partial}_i &= \frac{\partial x^k}{\partial \tilde{x}^i} \partial_k, \\
\partial_{ij} \to \tilde{\partial}_{ij} &= \frac{\partial^2 x^k}{\partial \tilde{x}^i \partial \tilde{x}^j} \partial_k + \frac{\partial x^k}{\partial \tilde{x}^i} \frac{\partial x^l}{\partial \tilde{x}^j} \partial_{kl},
\end{aligned} \tag{D.13}$$

and

$$
\begin{aligned}
d_2 x^i \;\rightarrow\; d_2 \tilde{x}^i \quad &= \frac{\partial \tilde{x}^i}{\partial x^k} d_2 x^k + \tfrac{1}{2} \frac{\partial^2 \tilde{x}^i}{\partial x^k \partial x^l} d[x^k, x^l] , \\
d[x^i, x^j] \;\rightarrow\; d[\tilde{x}^i, d\tilde{x}^j] &= \frac{\partial \tilde{x}^i}{\partial x^k} \frac{\partial \tilde{x}^j}{\partial x^l} d[x^k, x^l] .
\end{aligned}
\tag{D.14}
$$

However, if the manifold is equipped with an affine connection, one can construct covariant representations for second order vectors and forms. These are given by

$$
\begin{aligned}
\hat{v}^i &= v^i + \frac{1}{2} \Gamma^i_{jk} v^{jk} , \\
\hat{v}^{ij} &= v^{ij} , \\
\hat{\omega}_i &= \omega_i , \\
\hat{\omega}_{ij} &= \omega_{ij} - \Gamma^k_{ij} \omega_k .
\end{aligned}
\tag{D.15}
$$

Similarly, the covariant basis elements are

$$
\begin{aligned}
\hat{\partial}_i &= \partial_i , \\
\hat{\partial}_{ij} &= \partial_{ij} - \Gamma^k_{ij} \partial_k , \\
d_2 \hat{x}^i &= d_2 x^i + \frac{1}{2} \Gamma^i_{kl} d[x^k, x^l] , \\
d[\hat{x}^i, \hat{x}^j] &= d[x^i, x^j] .
\end{aligned}
\tag{D.16}
$$

These objects transform covariantly, i.e.

$$
\begin{aligned}
\hat{v}^i \;\rightarrow\; \tilde{\hat{v}}^i &= \hat{v}^k \frac{\partial \tilde{x}^i}{\partial x^k} , \\
\hat{\omega}_{ij} \;\rightarrow\; \tilde{\hat{\omega}}_{ij} &= \hat{\omega}_{kl} \frac{\partial x^k}{\partial \tilde{x}^i} \frac{\partial x^l}{\partial \tilde{x}^j} ,
\end{aligned}
\tag{D.17}
$$

and

$$
\begin{aligned}
\hat{\partial}_{ij} \;\rightarrow\; \tilde{\hat{\partial}}_{ij} &= \frac{\partial x^k}{\partial \tilde{x}^i} \frac{\partial x^l}{\partial \tilde{x}^j} \hat{\partial}_{kl} , \\
d_2 \hat{x}^i \;\rightarrow\; d_2 \tilde{\hat{x}}^i &= \frac{\partial \tilde{x}^i}{\partial x^k} d_2 \hat{x}^k .
\end{aligned}
\tag{D.18}
$$

We conclude this section by noting that second order geometry is a geometrical framework that generalizes structures from ordinary (first order) geometry to second order structures, as was done for vectors and forms in the preceding discussion. This can be done independently of any notions from stochastic analysis. However, the natural domain of application of second order geometry is stochastic calculus, as it provides an interpretation of the second order structures. It is straightforward to relate the previous general discussion on second order geometry to its application in Itô calculus, as discussed in Appendix B, by replacing

$$d_2 x \rightarrow d_\pm X \,,$$
$$d[x, x] \rightarrow \pm d[X, X] \,. \tag{D.19}$$

D.1 Maps Between First and Second Order Geometry

The preceding discussion can be reformulated in a coordinate independent way by constructing unique mappings between first and second order geometry, cf. e.g. Ref. [70].

For functions $f, g \in C^\infty(\mathcal{M})$ and linear operators L on $C^\infty(\mathcal{M})$, one can define *l'opérateur carré du champ* or the *squared field operator* by

$$\Gamma_L(f, g) := L(f\,g) - f\,L(g) - g\,L(f) \,. \tag{D.20}$$

Since $\Gamma_L(f, g) = 0$, if L is a derivation, this operator is a measure for how close the operator L is to being a derivation. Moreover, it provides an alternative definition of the quadratic variation, since

$$\Gamma_{d_2}(f, g) = d[f, g] \,. \tag{D.21}$$

Starting from first order vectors $A, B \in T\mathcal{M}$, one can construct second order vectors by taking their product $A B \in T_2\mathcal{M}$. Similarly, starting from first order forms, one can construct second order forms, using a map

$$\mathcal{H} : T^*\mathcal{M} \otimes T^*\mathcal{M} \rightarrow T_2^*\mathcal{M} \quad \text{s.t.} \quad \alpha \otimes \beta \mapsto \alpha \cdot \beta \,, \tag{D.22}$$

which satisfies the following properties:

$$\begin{aligned}
\langle \mathcal{H}(b), A\,B \rangle &= \tfrac{1}{2}\,[b(A, B) + b(B, A)] \quad \forall\, A, B \in T\mathcal{M},\ b \in T^2(T^*\mathcal{M})\,, \\
df \cdot dg &= \tfrac{1}{2}\,d[f, g] \qquad\qquad\qquad \forall\, f, g \in C^\infty(\mathcal{M}) \,.
\end{aligned} \tag{D.23}$$

Moreover, the adjoint $\mathcal{H}^* : T_2\mathcal{M} \rightarrow T\mathcal{M} \otimes T\mathcal{M}$ defines a map from second order vectors to symmetric $(2, 0)$-tensors.

Vice versa, any second order form can be projected onto a first order form in an intrinsic way using a map $\mathcal{P} : T_2^*\mathcal{M} \rightarrow T^*\mathcal{M}$, such that

$$\begin{aligned}
\mathcal{P}(d_2 f) &= df \quad \forall\, f \in C^\infty(\mathcal{M})\,, \\
\mathcal{P}(\alpha \cdot \beta) &= 0 \quad\ \forall\, \alpha, \beta \in T^*\mathcal{M} \,.
\end{aligned} \tag{D.24}$$

Alternatively, second order forms can be constructed using a linear map $\underline{d} : T^*\mathcal{M} \rightarrow T_2^*\mathcal{M}$, which satisfies the following properties

$$\begin{aligned}
\underline{d}(df) &= d_2 f & \forall f \in C^\infty(\mathcal{M})\,, \\
\underline{d}(f\,\alpha) &= df \cdot \alpha + f\,\underline{d}\alpha & \forall \alpha \in T^*\mathcal{M},\ f \in C^\infty(\mathcal{M})\,, \\
\mathcal{P}(\underline{d}\alpha) &= \alpha & \forall \alpha \in T^*\mathcal{M}\,, \\
\langle \underline{d}\alpha,\, AB - BA \rangle &= \langle \alpha,\, [A, B] \rangle & \forall \alpha \in T^*\mathcal{M},\ A, B \in T\mathcal{M}\,, \\
\langle \underline{d}\alpha,\, AB + BA \rangle &= A\langle \alpha, B \rangle + B\langle \alpha, A \rangle & \forall \alpha \in T^*\mathcal{M},\ A, B \in T\mathcal{M}\,.
\end{aligned} \tag{D.25}$$

Finally, using the affine connection $\Gamma : \mathfrak{X}(\mathcal{M}) \times \mathfrak{X}(\mathcal{M}) \to \mathfrak{X}(\mathcal{M})$. One can construct mappings $\mathcal{F} : T_2\mathcal{M} \to T\mathcal{M}$ and $\mathcal{G} : T^*\mathcal{M} \to T_2^*\mathcal{M}$ through the following relations

$$\begin{aligned}
(\mathcal{F}V)\,f &= V f - \langle \mathcal{H}\,\Gamma^*(df), V \rangle & \forall V \in T_2\mathcal{M},\ f \in C^\infty(\mathcal{M})\,, \\
\Gamma(A, B) &= A B f - F(A B)\,f & \forall A, B \in T\mathcal{M},\ f \in C^\infty(\mathcal{M})\,,
\end{aligned} \tag{D.26}$$

and

$$\begin{aligned}
\mathcal{G}(df) &= d_2 f - \mathcal{H}\,\Gamma^*(df) & \forall f \in C^\infty(\mathcal{M})\,, \\
\Gamma(A, B)\,f &= A B f - \langle \mathcal{G}(df), AB \rangle & \forall A, B \in T\mathcal{M},\ f \in C^\infty(\mathcal{M})\,.
\end{aligned} \tag{D.27}$$

$\mathcal{F}, \mathcal{G}$ then satisfy the following properties:

$$\begin{aligned}
\mathcal{F}(f\,V) &= f\,\mathcal{F}(V) & \forall V \in T_2\mathcal{M},\ f \in C^\infty(\mathcal{M})\,, \\
\mathcal{F}(A) &= A & \forall A \in T\mathcal{M} \subset T_2\mathcal{M}\,, \\
\mathcal{G}(f\,\alpha) &= f\,\mathcal{G}(\alpha) & \forall \alpha \in T^*\mathcal{M},\ f \in C^\infty(\mathcal{M})\,, \\
\mathcal{P}[\mathcal{G}(\alpha)] &= \alpha & \forall \alpha \in T^*\mathcal{M}\,, \\
\langle \alpha, \mathcal{F}(V) \rangle &= \langle \mathcal{G}(\alpha), V \rangle & \forall \alpha \in T^*\mathcal{M},\ V \in T_2\mathcal{M}\,.
\end{aligned} \tag{D.28}$$

We conclude this section by summarizing the action of the various maps in a local coordinate frame. Given two first order vector fields $A, B \in T\mathcal{M}$ their product is a second order vector given by

$$A B = \left[A^i\,\partial_i(B^j) \right] \partial_j + \left[A^i\,B^j \right] \partial_{ij}\,. \tag{D.29}$$

Given a second order form $\omega \in T_2^*\mathcal{M}$, the projection map acts as

$$\mathcal{P}(\omega) = \mathcal{P}\left(\omega_i\,d_2 x^i + \frac{1}{2}\,\omega_{ij}\,d[x^i, x^j] \right) = \omega_i\,dx^i\,. \tag{D.30}$$

Given a first order form $\alpha \in T^*\mathcal{M}$, the map $\underline{d}$ acts as

$$\underline{d}(\alpha) = \underline{d}\left(\alpha_i\,dx^i \right) = \alpha_i\,d_2 x^i + \frac{1}{2}\,\partial_j \alpha_i\,d[x^i, x^j]\,. \tag{D.31}$$

Given a first order form $\alpha \in T^*\mathcal{M}$, the map $\mathcal{G}$ acts as

$$\mathcal{G}(\alpha) = \mathcal{G}(\alpha_i\,dx^i) = \alpha_i \left(d_2 x^i + \frac{1}{2}\,\Gamma^i_{kl}\,d[x^k, x^l] \right) = \alpha_i\,d_2 \hat{x}^i\,. \tag{D.32}$$

Given two first order forms $\alpha, \beta \in T^*\mathcal{M}$, the map $\mathcal{H}$ acts as

$$\mathcal{H}(\alpha \otimes \beta) = \mathcal{H}(\alpha_i \, \beta_j \, dx^i \otimes dx^j) = \frac{1}{2} \, \alpha_i \, \beta_j \, d[x^i, x^j] \, . \tag{D.33}$$

Given a second order vector field $V \in T_2^*\mathcal{M}$, the map $\mathcal{F}$ acts as

$$\mathcal{F}(V) = \mathcal{G}\left(V^i \, \partial_i + \frac{1}{2} \, V^{ij} \, \partial_{ij} \right) = \left(V^i + \frac{1}{2} \, \Gamma_{kl}^i \, V^{kl} \right) \partial_i = \hat{V}^i \, \partial_i \, . \tag{D.34}$$

Given a second order vector field $V \in T_2^*\mathcal{M}$, the map $\mathcal{H}^*$ acts as

$$\mathcal{H}^*(V) = \mathcal{H}^*\left(V^i \, \partial_i + \frac{1}{2} \, V^{ij} \, \partial_{ij} \right) = \frac{1}{2} \, V^{ij} \, \partial_i \otimes \partial_j \, . \tag{D.35}$$

D.2   The Second Order Tangent Bundle

The (co)tangent bundle in first order geometry can be defined as follows.

Definition *(Tangent bundle)* The tangent bundle $(T\mathcal{M}, \tau_\mathcal{M}, \mathcal{M})$ is the fiber bundle with base space $\mathcal{M}$, projection $\tau_\mathcal{M} : T\mathcal{M} \to \mathcal{M}$, typical fiber $\mathbb{R}^n$ and structure group $GL(n, \mathbb{R})$ acting from the left.

Definition *(Cotangent bundle)* The cotangent bundle $(T^*\mathcal{M}, \tau_\mathcal{M}^*, \mathcal{M})$ is the fiber bundle dual to $(T\mathcal{M}, \tau_\mathcal{M}, \mathcal{M})$ with base space $\mathcal{M}$, projection $\tau_\mathcal{M}^* : T^*\mathcal{M} \to \mathcal{M}$, typical fiber $(\mathbb{R}^n)^*$ and structure group $GL(n, \mathbb{R})$ acting from the right.

These bundles can also be constructed as the bundle of tangent spaces at points $x \in \mathcal{M}$:

$$T\mathcal{M} = \bigsqcup_{x \in \mathcal{M}} T_x\mathcal{M} \, ,$$

$$T^*\mathcal{M} = \bigsqcup_{x \in \mathcal{M}} T_x^*\mathcal{M} \, . \tag{D.36}$$

Second order (co)tangent bundles can be defined in a similar way. Here, we quote the definitions given in Ref. [71].

Definition *(Itô group)* The Itô group G_I^n is the Cartesian product $GL(n, \mathbb{R}) \times \text{Lin}(\mathbb{R}^n \otimes \mathbb{R}^n, \mathbb{R}^n)$ equipped with the binary operation

$$(g', \kappa') \, (g, \kappa) = (g' \, g, \, g' \circ \kappa + \kappa' \circ (g \otimes g)) \tag{D.37}$$

for all $g, g' \in GL(n, \mathbb{R})$ and $\kappa, \kappa' \in \text{Lin}(\mathbb{R}^n \otimes \mathbb{R}^n, \mathbb{R}^n)$.

Definition (*Left action of the Itô group*) The left group action of G_I^n on $\mathbb{R}^n \times$ Sym$(T\mathbb{R}^n \otimes T\mathbb{R}^n)$ is defined by

$$(g, \kappa)(x, x_2) = (g\,x + \kappa\,x_2, (g \otimes g)\,x_2) \tag{D.38}$$

for all $(g, \kappa) \in G_I^n$, $x \in \mathbb{R}^n$ and $x_2 \in$ Sym$(T\mathbb{R}^n \otimes T\mathbb{R}^n)$.

Definition (*Right action of the Itô group*) The right group action of G_I^n on $(\mathbb{R}^n \times$ Sym$(T\mathbb{R}^n \otimes T\mathbb{R}^n))^*$ is given by

$$(p, p_2)(g, \kappa) = (g^*\,p, \; \kappa^*\,p + (g^* \otimes g^*)\,p_2) \tag{D.39}$$

for all $(g, \kappa) \in G_I^n$, $p \in (\mathbb{R}^n)^*$ and $p_2 \in$ Sym$(T\mathbb{R}^n \otimes T\mathbb{R}^n)^*$.

Definition (*Second order tangent bundle*) The second order tangent bundle $(T_2\mathcal{M}, \tau_\mathcal{M}^2, \mathcal{M})$ is the fiber bundle with base space $\mathcal{M}$, projection $\tau_\mathcal{M}^2 : T_2\mathcal{M} \to \mathcal{M}$, typical fiber $\mathbb{R}^n \times$ Sym$(T\mathbb{R}^n \otimes T\mathbb{R}^n)$ and structure group G_I^n acting from the left.

Definition (*Second order cotangent bundle*) The second order cotangent bundle $(T_2^*\mathcal{M}, \tau_\mathcal{M}^{2*}, \mathcal{M})$ is the fiber bundle dual to $(T_2\mathcal{M}, \tau_\mathcal{M}^2, \mathcal{M})$ with base space $\mathcal{M}$, projection $\tau_\mathcal{M}^{2*} : T_2^*\mathcal{M} \to \mathcal{M}$, typical fiber $(\mathbb{R}^n \times$ Sym$(T\mathbb{R}^n \otimes T\mathbb{R}^n))^*$ and structure group G_I^n acting from the right.

We note that these bundles can also be obtained from the second order (co)tangent spaces that were introduced earlier in this appendix, since

$$T_2\mathcal{M} = \bigsqcup_{x \in \mathcal{M}} T_{2,x}\mathcal{M},$$
$$T_2^*\mathcal{M} = \bigsqcup_{x \in \mathcal{M}} T_{2,x}^*\mathcal{M}. \tag{D.40}$$

D.3 Stochastic Integration on Manifolds

Integration on manifolds along deterministic trajectories is performed by forms $\omega \in T^*\mathcal{M}$. Using second order order geometry, this can be extended to stochastic integration on manifolds by mapping these first order forms to second order forms.

The Stratonovich integral is defined for any first order form $\omega \in T^*\mathcal{M}$ and is given by

$$\oint_{X_T} \omega = \int_T \omega_i(X_t)\, d_\circ X_t^i. \tag{D.41}$$

The right hand side can then be evaluated in a local coordinate chart using the definition of the Stratonovich integral on $\mathbb{R}^n$ given in Eq. (C.1). Alternatively, the Stratonovich integral can be defined as an integral over second order forms using the map $\underline{d}$ and the relation (C.6) between Itô and Stratonovich integrals on $\mathbb{R}^n$:

$$
\begin{aligned}
\fint_{X_T} \omega &:= \int_{X_T} \underline{d}_\pm(\omega) \\
&= \int_T \omega_i(X_t)\, d_\pm X_t^i \pm \frac{1}{2} \int_T \partial_j \omega_i(X_t)\, d[X^i, X^j]_t \\
&= \int_T \omega_i(X_t)\, d_\circ X_t^i .
\end{aligned}
\tag{D.42}
$$

The Itô integral is defined using the map $\mathcal{G}$, such that the forward Itô integral is given by

$$
\begin{aligned}
\underline{\int}_{X_T} \omega &:= \int_{X_T} \mathcal{G}_+(\omega) \\
&= \int_T \omega_i(X_t)\, d_+ \hat{X}_t^i \\
&= \int_T \omega_i(X_t)\, d_+ X_t^i + \frac{1}{2} \int_T \omega_i(X_t)\, \Gamma_{kl}^i(X_t)\, d[X^k, X^l]_t
\end{aligned}
\tag{D.43}
$$

and the backward Itô integral by

$$
\begin{aligned}
\overline{\int}_{X_T} \omega &:= \int_{X_T} \mathcal{G}_-(\omega) \\
&= \int_T \omega_i(X_t)\, d_- \hat{X}_t^i \\
&= \int_T \omega_i(X_t)\, d_- X_t^i - \frac{1}{2} \int_T \omega_i(X_t)\, \Gamma_{kl}^i(X_t)\, d[X^k, X^l]_t .
\end{aligned}
\tag{D.44}
$$

The right hand side of these expressions can be evaluated using the definitions of the Itô integral and the integral over quadratic variation on $\mathbb{R}^n$, given in Eqs. (C.2), (C.3) and (C.4).

The integral over the quadratic variation is defined as an integral over a $(0, 2)$-tensor $h \in T^2(T^*\mathcal{M})$ and is given by

$$
\begin{aligned}
\int_{X_T} h &= \int_{X_T} \mathcal{H}(h) \\
&= \frac{1}{2} \int_T h_{ij}(X_t)\, d[X^i, X^j]_t ,
\end{aligned}
\tag{D.45}
$$

where the right hand side can be evaluated using the definition of the integral over quadratic variation on $\mathbb{R}^n$, given in Eq. (C.4). We note that this integral is defined for any bilinear form h, but only the symmetric part of h contributes to the integral.

Finally, using the definition of the various integrals, one can derive a relation between the Stratonovich and the Itô integral. In a local coordinate frame, it is given by

$$\int_T \omega_i(X_t)\, d_\circ X_t^i = \int_T \omega_i(X_t)\, d_\pm X_t^i \pm \frac{1}{2} \int_T \partial_j \omega_i(X_t)\, d[X^i, X^j]_t$$

$$= \int_T \omega_i(X_t)\, d_\pm \hat{X}_t^i \pm \frac{1}{2} \int_T \nabla_j \omega_i(X_t)\, d[X^i, X^j]_t, \qquad (D.46)$$

which can be written in a coordinate independent way as

$$\oint_{X_T} \omega = \underline{\int}_{X_T} \omega + \int_{X_T} \nabla \omega$$

$$= \overline{\int}_{X_T} \omega - \int_{X_T} \nabla \omega. \qquad (D.47)$$

Appendix E
Construction of the Itô Lagrangian

In this appendix, we calculate the Itô Lagrangian $L^{\pm}(x, v_{\pm}, v_2, t)$ corresponding to the Stratonovich Lagrangian (3.37) on a pseudo-Riemannian manifold:

$$L^{\circ}(x, v_{\circ}, t) = \frac{m}{2} g_{ij}(x, t) \, v_{\circ}^i v_{\circ}^j + q \, A_i(x, t) \, v_{\circ}^i - \mathfrak{U}(x, t). \tag{E.1}$$

We do this by writing down the action for this Lagrangian

$$
\begin{aligned}
S_{\circ}(X) &= \mathbb{E}\left[\int_{t_0}^{t_f} L^{\circ}(X_t, V_{\circ,t}, t) \, dt\right] \\
&= \mathbb{E}\left[\int_{t_0}^{t_f} \left(\frac{m}{2} g_{ij}(X_t, t) \, V_{\circ,t}^i V_{\circ,t}^j + q \, A_i(X_t, t) \, V_{\circ,t}^i - \mathfrak{U}(X_t, t)\right) dt\right]
\end{aligned} \tag{E.2}
$$

and imposing that

$$S_{\circ}(X) = S_{\pm}(X) = \mathbb{E}\left[\int_{t_0}^{t_f} L^{\pm}(X_t, V_{\pm,t}, V_{2,t}, t) \, dt\right]. \tag{E.3}$$

Thus, we have to rewrite the Stratonovich velocity as an Itô velocity process. This can be done by evaluating the three terms in the action separately. For the third term, involving the scalar potential, this is straightforward, as it does not depend on the velocity process. Therefore, the third term is the same in the Itô and Stratonovich formulation.

For the second term, which is linear in velocity, we find

$$
\begin{aligned}
\mathbb{E}\left[\int_{t_0}^{t_f} A_i(X_t, t) V_{\circ,t}^i \, dt\right] &= \mathbb{E}\left[\int_{t_0}^{t_f} A_i(X_t, t) \, d_{\circ} X_t^i\right] \\
&= \mathbb{E}\left[\int_{t_0}^{t_f} \left(A_i(X_t, t) \, d_{\pm} X_t^i \pm \frac{1}{2} \partial_j A_i(X_t, t) \, d[X^i, X^j]_t\right)\right]
\end{aligned}
$$

F. Kuipers, *Stochastic Mechanics*, SpringerBriefs in Physics, https://doi.org/10.1007/978-3-031-31448-3

$$= \mathbb{E}\left[\int_{t_0}^{t_f}\left(A_i(X_t,t)\,V_{\pm,t}^i \pm \frac{1}{2}\partial_j A_i(X_t,t)\,V_{2,t}^{ij}\right)dt\right]$$

$$= \mathbb{E}\left[\int_{t_0}^{t_f}\left(A_i(X_t,t)\,\hat{V}_{\pm,t}^i \pm \frac{1}{2}\nabla_j A_i(X_t,t)\,V_{2,t}^{ij}\right)dt\right], \qquad (E.4)$$

where $\hat{V}_\pm^i = V_\pm^i \pm \frac{1}{2}\Gamma_{jk}^i\,V_2^{jk}$ is the covariant velocity process.

For the first term, which is quadratic in velocity, we find

$$\mathbb{E}\left[\int g_{ij}\,V_o^i V_o^j\,dt\right] = \mathbb{E}\left[\int g_{ij}\,V_o^i\,d_o X^j\right]$$

$$= \mathbb{E}\left[\int \delta_{ab}\,e_j^b\,V_o^a\,d_o X^j\right]$$

$$= \mathbb{E}\left[\int\left(\delta_{ab}\,e_j^b\,V_o^a\,d_\pm X^j \pm \frac{1}{2}\frac{\partial}{\partial x^k}\left(\delta_{ab}\,e_j^b\,V_o^a\right)d[X^j,X^k]\right.\right.$$

$$\left.\left.\pm\frac{1}{2}\frac{\partial}{\partial v^c}\left(\delta_{ab}\,e_j^b\,V_o^a\right)d[X^j,V_o^c]\right)\right]$$

$$= \mathbb{E}\left[\int\left(g_{ij}\,V_o^i\,d_\pm X^j \pm \frac{1}{2}\delta_{ab}\,\Gamma_{kj}^l\,e_l^b\,V_o^a\,d[X^j,X^k] \pm \frac{1}{2}\delta_{ab}\,e_j^b\,d[X^j,V_o^a]\right)\right]$$

$$= \mathbb{E}\left[\int\left(g_{ij}\,V_o^i\,d_\pm X^j \pm \frac{1}{2}g_{ij}\,\Gamma_{kl}^j\,V_o^i\,d[X^k,X^l] \pm \frac{1}{2}g_{ij}\,d[X^j,V_o^i]\right)\right]$$

$$= \mathbb{E}\left[\int\left(g_{ij}\,V_o^i\,\hat{V}_\pm^j\,dt \pm \frac{1}{2}g_{ij}\,d[X^i,V_o^j]\right)\right]. \qquad (E.5)$$

We can then repeat the same calculation for V_o^i, which yields

$$\mathbb{E}\left[\int g_{ij}\,V_o^i V_o^j\,dt\right] = \mathbb{E}\left[\int\left(g_{ij}\,\hat{V}_\pm^i\,\hat{V}_\pm^j\,dt \pm \frac{1}{2}g_{ij}\,d[X^i,V_o^j + \hat{V}_\pm^j]\right)\right]. \qquad (E.6)$$

An expression for the quadratic variation $[X,V_o]$ is given in Sect. 3.6, but we do not yet have an expression for $[X,\hat{V}_\pm]$. In order to obtain this expression, we rewrite the result of Eq. (E.5) using

$$V_o = \frac{1}{2}\left(\hat{V}_+ + \hat{V}_-\right),$$

$$V_\perp = \frac{1}{2}\left(\hat{V}_+ - \hat{V}_-\right). \qquad (E.7)$$

The first expression follows from the definition of the Stratonovich velocity and the second defines a velocity perpendicular to the Stratonovich velocity. We recalculate the quadratic term starting from the result in Eq. (E.5) and find

$$\mathbb{E}\left[\int g_{ij} V_\circ^i V_\circ^j \, dt\right] = \mathbb{E}\left[\int \left(g_{ij} V_\circ^i (V_\circ^j \pm \hat{V}_\perp^j) \, dt \pm \frac{1}{2} g_{ij} \, d[X^i, V_\circ^j]\right)\right]$$

$$= \mathbb{E}\left[\int \left(g_{ij} V_\circ^i \hat{V}_\pm^j \, dt \pm g_{ij} V_\circ^i \hat{V}_\perp^j \, dt \pm g_{ij} \, d[X^i, V_\circ^j]\right)\right],$$

(E.8)

where we repeated the calculation of Eq. (E.5) in the second line. Now, we must calculate the second term involving the perpendicular velocity. Repeating the calculation of Eq. (E.5), this term can be written as

$$\mathbb{E}\left[\int g_{ij} V_\circ^i \hat{V}_\perp^j \, dt\right] = \mathbb{E}\left[\int \left(g_{ij} \hat{V}_\pm^i \hat{V}_\perp^j \, dt \pm \frac{1}{2} g_{ij} \, d[X^i, \hat{V}_\perp^j]\right)\right]. \qquad (E.9)$$

When plugged into Eq. (E.8), this immediately yields our earlier result (E.6). However, this expression can also be evaluated in a different way, since, unlike the other velocities, the perpendicular velocity represents an acceleration. This can be seen by writing the expressions of the velocity in differential notation:

$$V_+ \, dt = X(t + dt) - X(t),$$
$$V_- \, dt = X(t) - X(t - dt),$$
$$V_\circ \, dt = \frac{1}{2}\left[X(t + dt) - X(t - dt)\right],$$
$$V_\perp \, dt = \frac{1}{2}\left[X(t + dt) - 2\,X(t) + X(t - dt)\right]. \qquad (E.10)$$

Therefore, $d[X, V_\perp]$ can be reduced to the quadratic variation $d[X, X]$ by improving Itô's lemma by one order in the Taylor expansion. This yields

$$\mathbb{E}\left[\int g_{ij} V_\circ^i \hat{V}_\perp^j \, dt\right] = \mathbb{E}\left[\int \left(g_{ij} \hat{V}_\perp^j \, d_\circ X^i\right)\right]$$

$$= \mathbb{E}\left[\int \left(\delta_{ab} \, e_i^a \, V_\perp^b \, d_\circ X^i\right)\right]$$

$$= \mathbb{E}\left[\int \left(\delta_{ab} \, e_i^a \, V_\perp^b \, d_\pm X^i \pm \frac{1}{2} \frac{\partial}{\partial x^k}\left(\delta_{ab} \, e_i^a\right) V_\perp^b \, d[X^i, X^k]\right.\right.$$

$$\pm \frac{1}{6} \frac{\partial^2}{\partial x^l \partial x^k}\left(\delta_{ab} \, e_i^a\right) e_j^b \, d[X^j, X^l]\, d[X^i, X^k]$$

$$\left.\left.\mp \frac{1}{6} \delta_{ab} \, e_i^a \frac{\partial^2}{\partial x^l \partial x^k}\left(e_j^b\right) d[X^j, X^l]\, d[X^i, X^k]\right)\right]$$

$$= \mathbb{E}\left[\int \left(g_{ij} \hat{V}_\perp^j \, d_\pm X^i \pm \frac{1}{2} g_{ij} \hat{V}_\perp^j \, \Gamma_{kl}^i \, d[X^k, X^l]\right.\right.$$

$$\pm \frac{1}{6} g_{mj} \left(\partial_l \Gamma_{ki}^m + \Gamma_{ln}^m \Gamma_{ki}^n\right) d[X^j, X^l]\, d[X^i, X^k]$$

$$\mp \frac{1}{6} g_{mi} \left(\partial_l \Gamma^m_{kj} + \Gamma^m_{ln} \Gamma^n_{kj} \right) d[X^j, X^l] d[X^i, X^k] \right) \right]$$

$$= \mathbb{E}\left[\int \left(g_{ij} \hat{V}^j_\perp d_\pm \hat{X}^i \pm \frac{1}{6} g_{im} \mathcal{R}^m{}_{jkl} d[X^i, X^k] d[X^j, X^l] \right) \right]$$

$$= \mathbb{E}\left[\int \left(g_{ij} \hat{V}^i_\pm \hat{V}^j_\perp \pm \frac{1}{6} \mathcal{R}_{ijkl} V^{ik}_2 V^{jl}_2 \right) dt \right]. \tag{E.11}$$

This result can be plugged into Eq. (E.8), which yields the result for the quadratic term

$$\mathbb{E}\left[\int g_{ij} V^i_\circ V^j_\circ \, dt \right] = \mathbb{E}\left[\int \left(g_{ij} \hat{V}^i_\pm \hat{V}^j_\pm \, dt + \frac{1}{6} \mathcal{R}_{ijkl} V^{ik}_2 V^{jl}_2 \, dt \pm g_{ij} d[X^i, V^j_\circ] \right) \right]. \tag{E.12}$$

Moreover, by comparing to Eqs. (E.6) and (E.9), we find

$$d[X^i, \hat{V}^j_\pm] = d[X^i, V^j_\circ] \pm d[X^i, \hat{V}^j_\perp]$$

$$= d[X^i, V^j_\circ] \pm \frac{1}{3} \mathcal{R}_{ijkl} V^{jl}_2 d[X^i, X^k]. \tag{E.13}$$

Since all terms appearing in the Stratonovich Lagrangian have been rewritten into an Itô formulation, we can finally impose Eq. (E.3). This yields

$$L^\pm(x, v_\pm, v_2, t) = L^\pm_0(x, v_\pm, v_2, t) \pm L_\infty(x, v_\circ), \tag{E.14}$$

where the finite part is given by

$$L^\pm_0(x, v_\pm, v_2, t) = \frac{m}{2} g_{ij} \left(v^i_\pm \pm \frac{1}{2} \Gamma^i_{kl} v^{kl}_2 \right) \left(v^j_\pm \pm \frac{1}{2} \Gamma^j_{kl} v^{kl}_2 \right) + \frac{m}{12} \mathcal{R}_{ijkl} v^{ik}_2 v^{jl}_2$$

$$+ q A_i v^i_\pm \pm \frac{q}{2} v^{ij}_2 \partial_j A_i - \mathfrak{U}, \tag{E.15}$$

which can be rewritten into an explicitly covariant form as

$$L^\pm_0(x, \hat{v}_\pm, v_2, t) = \frac{m}{2} g_{ij} \hat{v}^i_\pm \hat{v}^j_\pm + \frac{m}{12} \mathcal{R}_{ijkl} v^{ik}_2 v^{jl}_2 + q A_i \hat{v}^i_\pm \pm \frac{q}{2} v^{ij}_2 \nabla_j A_i - \mathfrak{U}. \tag{E.16}$$

Furthermore, the divergent part L_∞ is defined by the integral condition

$$\mathbb{E}\left[\int L_\infty(x, v_\circ) \, dt \right] = \mathbb{E}\left[\int \frac{m}{2} g_{ij} d[x^i, v^j_\circ] \right]. \tag{E.17}$$

Appendix F
Stochastic Variational Calculus

F.1 Stratonovich-Euler-Lagrange Equations

We consider a set $\mathcal{T} = [t_0, t_f]$, a pseudo-Riemannian manifold $\mathcal{M}$, a Stratonovich tangent bundle $T_\circ \mathcal{M}$, a Stratonovich Lagrangian $L^\circ(x, v_\circ, t) : T_\circ \mathcal{M} \times \mathcal{T} \to \mathbb{R}$ and the action

$$S(X) = \mathbb{E}\left[\int_{t_0}^{t_f} L^\circ(X_t, V_{\circ,t}, t)\, dt\right], \tag{F.1}$$

where X is a continuous semi-martingale process on $\mathcal{M}$ with a Doob-Meyer decomposition of the form

$$X_t^i = C_t^i + e_a^i(X_t)\, M^a. \tag{F.2}$$

We can vary this action with respect to another process δX that is stochastically independent, i.e. $\delta M \perp\!\!\!\perp M$, and satisfies the boundary conditions $\delta X_{t_0} = \delta X_{t_f} = 0$. We find

$$
\begin{aligned}
\delta S(X) &= S(X + \delta X) - S(X) \\
&= \mathbb{E}\left[\int_{t_0}^{t_f} \left(L^\circ(X_t + \delta X_t, V_{\circ,t} + \delta V_{\circ,t}, t) - L^\circ(X_t, V_{\circ,t}, t)\right) dt\right] \\
&= \mathbb{E}\left[\int_{t_0}^{t_f} \left(\frac{\partial L^\circ}{\partial x^i}\, \delta X_t^i + \frac{\partial L^\circ}{\partial v_\circ^i}\, \delta V_{\circ,t}^i + \mathcal{O}(\delta X_t^2)\right) dt\right] \\
&= \mathbb{E}\left[\int_{t_0}^{t_f} \left(\frac{\partial L^\circ}{\partial x^i}\, \delta X_t^i\, dt + \frac{\partial L^\circ}{\partial v_\circ^i}\, d_\circ \delta X_t^i\right)\right] + \mathcal{O}||\delta X||^2 \\
&= \mathbb{E}\left[\frac{\partial L^\circ}{\partial v_\circ^i}\, \delta X_t^i\Big|_{t_0}^{t_f} + \int_{t_0}^{t_f} \delta X_t^i \left(\frac{\partial L^\circ}{\partial x^i}\, dt - d_\circ \frac{\partial L^\circ}{\partial v_\circ^i}\right)\right] + \mathcal{O}||\delta X||^2 \\
&= \mathbb{E}\left[\int_{t_0}^{t_f} \delta X_t^i \left(\frac{\partial L^\circ}{\partial x^i}\, dt - d_\circ \frac{\partial L^\circ}{\partial v_\circ^i}\right)\right] + \mathcal{O}||\delta X||^2,
\end{aligned}
\tag{F.3}
$$

© The Editor(s) (if applicable) and The Author(s), under exclusive license to Springer Nature Switzerland AG 2023
F. Kuipers, *Stochastic Mechanics*, SpringerBriefs in Physics,
https://doi.org/10.1007/978-3-031-31448-3

where we used the Stratonovich integration by parts formula in the fifth line and $\delta X_{t_0} = \delta X_{t_f} = 0$ in the sixth line.

By taking the limit $||\delta X|| \to 0$, we find

$$\frac{\delta S(X)}{\delta X^i} = \mathbb{E}\left[\int_{t_0}^{t_f} \left(\frac{\partial L^\circ}{\partial x^i}\, dt - d_\circ \frac{\partial L^\circ}{\partial v_\circ^i}\right)\right]. \tag{F.4}$$

If we impose $\frac{\delta S}{\delta X} = 0$, this yields the Stratonovich-Euler-Lagrange equations

$$\mathbb{E}\left[\int_{t_0}^{t_f} d_\circ \frac{\partial}{\partial v_\circ^i} L^\circ(X_t, V_{\circ,t}, t)\right] = \mathbb{E}\left[\int_{t_0}^{t_f} \frac{\partial}{\partial x^i} L^\circ(X_t, V_{\circ,t}, t)\, dt\right], \tag{F.5}$$

which can be written in a differential notation as

$$d_\circ \frac{\partial L^\circ}{\partial v_\circ^i} = \frac{\partial L^\circ}{\partial x^i}\, dt. \tag{F.6}$$

F.2   Itô-Euler-Lagrange Equations

We consider a set $\mathcal{T} = [t_0, t_f]$, a pseudo-Riemannian manifold $\mathcal{M}$, Itô tangent bundles $T_\pm \mathcal{M}$, Itô Lagrangians $L^\pm(x, v_\pm, v_2, t) : T_\pm \mathcal{M} \times \mathcal{T} \to \mathbb{R}$ and the action

$$S(X) = \mathbb{E}\left[\int_{t_0}^{t_f} L^\pm(X_t, V_{\pm,t}, V_{2,t}, t)\, dt\right]. \tag{F.7}$$

where X is a continuous semi-martingale process on $\mathcal{M}$ with a Doob-Meyer decomposition of the form

$$X_t^i = C_t^i + e_a^i(X_t) M_t^a. \tag{F.8}$$

We can vary this action with respect to another process δX that is stochastically independent, i.e. $\delta M \perp\!\!\!\perp M$, and satisfies the boundary conditions $\delta X_{t_0} = \delta X_{t_f} = 0$. We find

$$\delta S(X) = S(X + \delta X) - S(X)$$

$$= \mathbb{E}\left[\int_{t_0}^{t_f} \left(L^\pm(X_t + \delta X_t, V_{\pm,t} + \delta V_{\pm,t}, V_{2,t} + \delta V_{2,t}, t) - L^\pm(X_t, V_{\pm,t}, V_{2,t}, t)\right) dt\right]$$

$$= \mathbb{E}\left[\int_{t_0}^{t_f} \left(\frac{\partial L^\pm}{\partial x^i} \delta X_t^i + \frac{\partial L^\pm}{\partial v_\pm^i} \delta V_{\pm,t}^i + \frac{\partial L^\pm}{\partial v_2^{ij}} \delta V_{2,t}^{ij} + \mathcal{O}(\delta X_t^2)\right) dt\right]$$

$$= \mathbb{E}\left[\int_{t_0}^{t_f} \left(\frac{\partial L^\pm}{\partial x^i} \delta X_t^i\, dt + \frac{\partial L^\pm}{\partial v_\pm^i} d_\pm \delta X_t^i + \frac{\partial L^\pm}{\partial v_2^{ij}} d[\delta X^i, X^j]_t + \frac{\partial L^\pm}{\partial v_2^{ij}} d[X^i, \delta X^j]_t\right)\right]$$

$$+ \mathcal{O}||\delta X||^2$$

$$= \mathbb{E}\left[\frac{\partial L^{\pm}}{\partial v_{\pm}^i}\,\delta X_t^i\Big|_{t_0}^{t_f} + \int_{t_0}^{t_f}\delta X_t^i\left(\frac{\partial L^{\pm}}{\partial x^i}\,dt - d_{\pm}\frac{\partial L^{\pm}}{\partial v_{\pm}^i}\right)\mp d\left[\frac{\partial L^{\pm}}{\partial v_{\pm}^i},\delta X^i\right]_t\right]$$

$$+ \mathbb{E}\left[\int_{t_0}^{t_f}\left(\frac{\partial L^{\pm}}{\partial v_2^{ij}}+\frac{\partial L^{\pm}}{\partial v_2^{ji}}\right)d[X^j,\delta X^i]_t\right] + \mathcal{O}\|\delta X\|^2$$

$$= \mathbb{E}\left[\int_{t_0}^{t_f}\left\{\delta X_t^i\left(\frac{\partial L^{\pm}}{\partial x^i}\,dt - d_{\pm}\frac{\partial L^{\pm}}{\partial v_{\pm}^i}\right)+\left(\frac{\partial L^{\pm}}{\partial v_2^{ij}}+\frac{\partial L^{\pm}}{\partial v_2^{ji}}\right)d[X^j,\delta X^i]_t\right.\right.$$

$$\left.\left.\mp\frac{\partial^2 L^{\pm}}{\partial x^j\partial v_{\pm}^i}\,d\left[X^j,\delta X^i\right]_t \mp\frac{\partial^2 L^{\pm}}{\partial v_{\pm}^j\partial v_{\pm}^i}\,d\left[V_{\pm}^j,\delta X^i\right]_t \mp\frac{\partial^2 L^{\pm}}{\partial v_2^{jk}\partial v_{\pm}^i}\,d\left[V_2^{jk},\delta X^i\right]_t\right\}\right]$$

$$+ \mathcal{O}\|\delta X\|^2,\tag{F.9}$$

where we used the Itô integraton by parts formula in the fifth line and $\delta X_{t_0} = \delta X_{t_f} = 0$ in the sixth line.

Since we have not yet factorized δX^i, we must further evaluate the quadratic variation containing δX^i. We find

$$d[X^j,\delta X^i] = e_a^j\,\delta e_b^i\,d[M^a,M^b] + e_a^j\,e_b^i\,d[M^a,\delta M^b] + \mathcal{O}(\delta X^2) + o(dt)$$

$$= -e_a^j\,\Gamma_{kl}^i\,e_b^l\,\delta X^k\,d[M^a,M^b] + \mathcal{O}(\delta X^2) + o(dt)$$

$$= -\Gamma_{kl}^i\,\delta X^k\,d[X^j,X^l] + \mathcal{O}(\delta X^2) + o(dt),\tag{F.10}$$

where we used the Doob-Meyer decomposition (F.8) in the first and last line, and the stochastic independence $\delta M \perp\!\!\!\perp M$, which implies $d[M^a,\delta M^b] = 0$, in the second line. By a similar calculation we find

$$d[V_{\pm}^j,\delta X^i] = -\Gamma_{kl}^i\,\delta X^k\,d[V_{\pm}^j,X^l] + \mathcal{O}(\delta X^2) + o(dt),\tag{F.11}$$

$$d[V_2^{jk},\delta X^i] = -\Gamma_{lm}^i\,\delta X^m\,d[V_2^{jk},X^l] + \mathcal{O}(\delta X^2) + o(dt).\tag{F.12}$$

These results can be plugged into the variation of the action. Then, in the limit $\|\delta X\| \to 0$, we obtain

$$\frac{\delta S(X)}{\delta X^i} = \mathbb{E}\left[\int_{t_0}^{t_f}\left(\frac{\partial L^{\pm}}{\partial x^i}\,dt - d_{\pm}\frac{\partial L^{\pm}}{\partial v_{\pm}^i} - \Gamma_{ij}^k\left(\frac{\partial L^{\pm}}{\partial v_2^{kl}}+\frac{\partial L^{\pm}}{\partial v_2^{lk}}\right)d[X^j,X^l]_t\right.\right.$$

$$\pm\Gamma_{ij}^k\,\frac{\partial^2 L^{\pm}}{\partial x^l\partial v_{\pm}^i}\,d[X^j,X^l]_t \pm\Gamma_{ij}^k\,\frac{\partial^2 L^{\pm}}{\partial v_{\pm}^l\partial v_{\pm}^k}\,d[X^j,V_{\pm}^l]_t$$

$$\left.\left.\pm\Gamma_{ij}^k\,\frac{\partial^2 L^{\pm}}{\partial v_2^{lm}\partial v_{\pm}^k}\,d[X^j,V_2^{lm}]_t\right)\right].\tag{F.13}$$

If we impose $\frac{\delta S}{\delta X} = 0$, this yields the Itô-Euler-Lagrange equations. These can be written in differential notation as

$$d_{\pm} \frac{\partial L^{\pm}}{\partial v_{\pm}^i} = \frac{\partial L^{\pm}}{\partial x^i} dt - \Gamma_{ij}^k \left(\frac{\partial L^{\pm}}{\partial v_2^{kl}} + \frac{\partial L^{\pm}}{\partial v_2^{lk}} \right) d[X^j, X^l]_t \pm \Gamma_{ij}^k \frac{\partial^2 L^{\pm}}{\partial x^l \partial v_{\pm}^k} d[X^j, X^l]_t$$

$$\pm \Gamma_{ij}^k \frac{\partial^2 L^{\pm}}{\partial v_{\pm}^l \partial v_{\pm}^k} d[X^j, V_{\pm}^l]_t \pm \Gamma_{ij}^k \frac{\partial^2 L^{\pm}}{\partial v_2^{lm} \partial v_{\pm}^k} d[X^j, V_2^{lm}]_t . \tag{F.14}$$

F.3   Hamilton-Jacobi-Bellman Equations

We consider a set $\mathcal{T} = [t_0, t_f]$, a pseudo-Riemannian manifold $\mathcal{M}$, Itô tangent bundles $T_{\pm}\mathcal{M}$ and Itô Lagrangians $L^{\pm}(x, v_{\pm}, v_2, \varepsilon, t) : T_{\pm}\mathcal{M} \times \mathbb{R} \times \mathcal{T} \to \mathbb{R}$, where $\varepsilon \in \mathbb{R}$ is a Lagrange multiplier. For L^+, we construct the principal function

$$S^+(x, \varepsilon, t) = S^+(x, \varepsilon, t; x_f, \varepsilon_f, t_f)$$

$$= -\mathbb{E} \left[\int_t^{t_f} L^+(X_s, V_{+,s}, V_{2,s}, \mathcal{E}_s, s) \, ds \, \Big| \, X_t = x, X_{t_f} = x_f, \mathcal{E}_t = \varepsilon, \mathcal{E}_{t_f} = \varepsilon_f \right], \tag{F.15}$$

where $(X_s, V_s, \mathcal{E}_s)$ is a solution of the Itô-Euler-Lagrange equations passing through (x, ε, t) and $(x_f, \varepsilon_f, t_f)$. For L^-, we construct the principal function

$$S^-(x, \varepsilon, t) = S^-(x, \varepsilon, t; x_0, \varepsilon_0, t_0)$$

$$= \mathbb{E} \left[\int_{t_0}^t L^-(X_s, V_{-,s}, V_{2,s}, \mathcal{E}_s, s) \, ds \, \Big| \, X_t = x, X_{t_0} = x_0, \mathcal{E}_t = \varepsilon, \mathcal{E}_{t_0} = \varepsilon_0 \right], \tag{F.16}$$

where $(X_s, V_s, \mathcal{E}_s)$ is a solution of the Itô-Euler-Lagrange equations passing through $(x_0, \varepsilon_0, t_0)$ and (x, ε, t).

We can vary these principal functions with respect to their end point (x, ε, t). For S^+, we obtain

$$\delta S^+(x, \varepsilon, t) = S^+(x + \delta x, \varepsilon, t) - S^+(x, \varepsilon, t)$$

$$= -\mathbb{E} \left\{ \mathbb{E} \left[\int_t^{t_f} L^+ \, ds \, \Big| \, X_t = x + \delta x \right] - \mathbb{E} \left[\int_t^{t_f} L^+ \, ds \, \Big| \, X_t = x \right] \Big| X_{t_f} = x_f \right\}$$

$$= -\mathbb{E} \left[\int_t^{t_f} (L_{X+\delta X}^+ - L_X^+) \, ds \, \Big| \, X_t = x, X_{t_f} = x_f, \delta X_t = \delta x, \delta X_{t_f} = 0 \right]$$

$$= -\mathbb{E} \left[\int_t^{t_f} \left(\frac{\partial L^+}{\partial x^i} \delta X_s^i + \frac{\partial L^+}{\partial v_+^i} \delta V_{+,s}^i \right. \right.$$

$$\left. \left. + \frac{\partial L^+}{\partial v_2^{ij}} \delta V_{2,s}^{ij} + \mathcal{O}(\delta X_s^2) \right) ds \, \Big| \, X_t, X_{t_f}, \delta X_{t_f}, \delta X_t \right]$$

$$= -\mathbb{E} \left[\int_t^{t_f} \left(\delta X_s^i \, d_+ \frac{\partial L^+}{\partial v_+^i} + \frac{\partial L^+}{\partial v_+^i} d_+ \delta X_s^i \right. \right.$$

$$\left. \left. + d \left[\frac{\partial L^+}{\partial v_+^i}, \delta X^i \right]_s \right) \Big| \, X_t, X_{t_f}, \delta X_{t_f}, \delta X_t \right] + \mathcal{O}\|\delta x\|^2$$

$$= -\mathbb{E}\left[\int_t^{t_f} d_+\left(\frac{\partial L^+}{\partial v_+^i}\,\delta X_s^i\right)\Bigg| X_t, X_{t_f}, \delta X_{t_f}, \delta X_t\right] + \mathcal{O}||\delta x||^2$$

$$= -\mathbb{E}\left[\frac{\partial L^+}{\partial v_+^i}\,\delta X_s^i\,\bigg|_t^{t_f}\,\bigg| X_t = x, \delta X_t = \delta x, X_{t_f} = x_f, \delta X_{t_f} = 0\right] + \mathcal{O}||\delta x||^2$$

$$= \mathbb{E}\left[\frac{\partial}{\partial v_+^i}L^+(X_t, V_{+,t}, V_{2,t}, \mathcal{E}_t, t)\,\bigg| X_t = x, \mathcal{E}_t = \varepsilon\right]\delta x^i + \mathcal{O}||\delta x||^2, \qquad \text{(F.17)}$$

where we suppressed the dependence on the initial and final condition of the Lagrange multiplier $\mathcal{E}_t$ in the first seven lines. Moreover, we used the Itô-Euler-Lagrange equation in the fifth line and the Itô integration by parts formula in the sixth line.

By a similar calculation, we find for S^-

$$\delta S^-(x, \varepsilon, t) = \mathbb{E}\left[\frac{\partial}{\partial v_-^i}L^-(X_t, V_{-,t}, V_{2,t}, \mathcal{E}_t, t)\,\bigg| X_t = x, \mathcal{E}_t = \varepsilon\right]\delta x^i + \mathcal{O}||\delta x||^2.$$

$$\text{(F.18)}$$

After taking the limit $||\delta x|| \to 0$, we obtain the first Hamilton-Jacobi equation

$$\frac{\partial}{\partial x^i}S^\pm(X_t, \mathcal{E}_t, t) = p_i^\pm(X_t, \mathcal{E}_t, t) \qquad \text{(F.19)}$$

with

$$p_i^\pm(X_t, \mathcal{E}_t, t) = \mathbb{E}\left[\frac{\partial L^\pm}{\partial v_\pm^i}\,\bigg| X_t, \mathcal{E}_t\right]. \qquad \text{(F.20)}$$

The second Hamilton-Jacobi equation can be obtained by applying Itô's lemma to $S^\pm$, which yields

$$d_\pm S^\pm(X_t, \mathcal{E}_t, t) = \frac{\partial S^\pm}{\partial t}\,dt + \frac{\partial S^\pm}{\partial \varepsilon}\,d\mathcal{E}_t + \frac{\partial S^\pm}{\partial x^i}\,d_\pm X_t^i \pm \frac{1}{2}\frac{\partial^2 S^\pm}{\partial x^j \partial x^i}\,d[X^i, X^j]_t$$

$$= \frac{\partial S^\pm}{\partial t}\,dt + \frac{\partial S^\pm}{\partial \varepsilon}\frac{d\mathcal{E}_t}{dt}\,dt + p_i^\pm\,d_\pm X_t^i \pm \frac{1}{2}\frac{\partial p_i^\pm}{\partial x^j}\,d[X^i, X^j]_t,$$

$$\text{(F.21)}$$

where we used that $\mathcal{E}_t$ is a deterministic process and the first Hamilton-Jacobi equation in the second line. By taking a conditional expectation of this expression, we find

$$\mathbb{E}\left[d_\pm S^\pm(X_t, \mathcal{E}_t, t)\,\bigg| X_t, \mathcal{E}_t\right] = \left[\frac{\partial S^\pm}{\partial t} + \frac{\partial S^\pm}{\partial \varepsilon}\frac{d\mathcal{E}_t}{dt} + p_i^\pm v_\pm^i \pm \frac{1}{2}v_2^{ij}\frac{\partial p_i^\pm}{\partial x^j}\right]dt. \qquad \text{(F.22)}$$

Moreover, using the definition of Hamilton's principal function, given in Eqs. (F.15) and (F.16), we obtain

$$\mathbb{E}\left[d_\pm S^\pm \middle| X_t, \mathcal{E}_t\right] = L^\pm \left[X_t, v_\pm(X_t, \mathcal{E}_t, t), v_2(X_t, \mathcal{E}_t, t), \mathcal{E}_t, t\right] dt$$

$$= \left\{L_0^\pm \left[X_t, v_\pm, v_2, \mathcal{E}_t, t\right] \pm L_\infty \left[X_t, v_o, \mathcal{E}_t\right]\right\} dt . \qquad \text{(F.23)}$$

Then, comparing Eqs. (F.22) and (F.23) at finite order yields

$$\frac{\partial}{\partial t} S(X_t, \mathcal{E}_t, t) + \frac{d\mathcal{E}_t}{dt} \frac{\partial}{\partial \varepsilon} S(X_t, \mathcal{E}_t, t) = -H_0^\pm \left[X_t, p^\pm(X_t, \mathcal{E}_t, t), \partial p^\pm(X_t, \mathcal{E}_t, t), \mathcal{E}_t, t\right]$$

$$\text{(F.24)}$$

with the Hamiltonian given by

$$H_0^\pm(x, p^\pm, \partial p^\pm, \varepsilon, t) = p_i^\pm v_\pm^i \pm \frac{1}{2} v_2^{ij} \, \partial_j p_i^\pm - L_0^\pm(x, v_\pm, v_2, \varepsilon, t)$$

$$= p_i^\pm \hat{v}_\pm^i \pm \frac{1}{2} v_2^{ij} \, \nabla_j p_i^\pm - L_0^\pm(x, v_\pm, v_2, \varepsilon, t) . \qquad \text{(F.25)}$$

Furthermore, comparing the divergent parts of Eqs. (F.22) and (F.23) yields an integral constraint

$$\mathbb{E}\left[\oint L_\infty(X_s, V_{o,s}, \mathcal{E}_s) \, ds \, \middle| \, X_t, \mathcal{E}_t\right] = \pm \oint \left(p_i^\pm \hat{v}_\pm^i \pm \frac{1}{2} v_2^{ij} \, \nabla_j p_i^\pm\right) dt . \qquad \text{(F.26)}$$

References

1. E. Nelson, *Dynamical Theories of Brownian Motion* (Princeton University Press, 1967)
2. R. Brown, A brief account of microscopical observations made in the months of June, July, and August, 1827, on the particles contained in the pollen of plants; and on the general existence of active molecules in organic and inorganic bodies. Philos. Mag. N. S. **4** (1828)
3. A. Einstein, Über die von der molekularkinetischen Theorie der Wärme geforderte Bewegung von in ruhenden Flässigkeiten suspendierten Teilchen. Annalen der Physik **322**(8), 549–560 (1905)
4. M.M. Smoluchowski, Sur le chemin moyen parcouru par les molécules d'un gaz et sur son rapport avec la théorie de la diffusion. Bull. Int. de l'Académie des Sci. de Cracovie **202** (1906)
5. J. Perrin, Brownian movement and molecular reality. Ann. de Chimie et de Physique 8 me Ser. (1909)
6. G.E. Uhlenbeck, L.S. Ornstein, On the theory of Brownian motion. Phys. Rev. **36**, 823–841 (1930)
7. P. Langevin, Sur la théorie du mouvement brownien. C. R. Acad. Sci. Paris. **146**, 530–533 (1908)
8. N. Wiener, Differential space. J. Math. Phys. **2**, 131–174 (1923)
9. A. Kolmogorov, *Grundbegriffe der Wahrscheinlichkeitsrechnung* (Springer, 1933)
10. J.L. Doob, The brownian movement and stochastic equations. Ann. Math. **43**(2), 351–369 (1942)
11. P. Lévy, *Processus stochastiques et mouvement brownien* (Gauthier-Villars, 1948)
12. K. Itô, Stochastic Integral. Proc. Imp. Acad. Tokyo **20**, 519–524 (1944)
13. D. Fisk, Quasimartingales. Trans. Am. Math. Soc. **120**, 369–389 (1965)
14. R.L. Stratonovich, *Conditional Markov Processes and Their Application to the Theory of Optimal Control* (Moscow University Press, Izd, 1966)
15. M. Planck, Über das Gesetz der Energieverteilung im Normalspektrum. Ann. Phys. **4**, 553 (1901)
16. A. Einstein, Über einen der Erzeugung und Verwandlung des Lichtes betreffenden heuristischen Gesichtspunkt. Ann. Phys. **17**, 132–148 (1905)
17. P.A.M. Dirac, *The Principles of Quantum Mechanics* (Oxford University Press, 1930)
18. J. von Neumann, *Mathematische Grundlagen der Quantenmechanik* (Springer, Berlin, 1932)
19. A.S. Wightman, Fields as operator-valued distributions in relativistic quantum theory. Arkiv f. Fysik, Kungl. Svenska Vetenskapsak. **28**, 129–189 (1964)

© The Editor(s) (if applicable) and The Author(s), under exclusive license to Springer Nature Switzerland AG 2023
F. Kuipers, *Stochastic Mechanics*, SpringerBriefs in Physics,
https://doi.org/10.1007/978-3-031-31448-3

20. R.P. Feynman, Space-time approach to nonrelativistic quantum mechanics. Rev. Mod. Phys. **20**, 367–387 (1948)
21. E. Nelson, Derivation of the Schrodinger equation from Newtonian mechanics. Phys. Rev. **150**, 1079–1085 (1966)
22. E. Nelson, *Quantum Fluctuations* (Princeton University Press, 1985)
23. A. Kolmogorov, Über die analytischen Methoden in der Wahrscheinlichkeitsrechnung. Mathematische Ann. **10**, 415–458 (1931)
24. M. Kac, On distribution of certain wiener functionals. Trans. Am. Math. Soc. **65**, 1–13 (1949)
25. G.C. Wick, Properties of bethe-salpeter wave functions. Phys. Rev. **96**, 1124–1134 (1954)
26. S.A. Albeverio, R.J. Høegh-Krohn, S. Mazzucchi, *Mathematical Theory of Feynman Path Integrals*. Lecture Notes in Mathematics, vol. 523 (Springer, 2008)
27. J. Schwinger, Four-dimensional Euclidean formulation of quantum field theory, in *Proceedings of the 8th Annual International Conference on High Energy Physics*, pp. 134–140 (1958)
28. E. Nelson, Construction of quantum fields from Markoff fields. J. Funct. Anal. **12**, 1, 97 (1973)
29. K. Osterwalder, R. Schrader, Axiom for Euclidean Green's functions. Commun. Math. Phys. **31**, 83–112 (1973)
30. K. Osterwalder, R. Schrader, Axioms for Euclidean Green's functions. 2. Commun. Math. Phys. **42**, 281 (1975)
31. J. Glimm, A.M. Jaffe, *Quantum Physics: a Functional Integral Point of View* (Springer, New York, 1987)
32. G. Parisi, Y.S. Wu, Perturbation theory without gauge fixing. Sci. Sin. **24**, 483 (1981)
33. P.H. Damgaard, H. Huffel, Stochastic quantization. Phys. Rept. **152**, 227 (1987)
34. R.H. Cameron, A family of integrals serving to connect the wiener and feynman integrals. J. Math. Phys. **39**, 126–140 (1960)
35. Yu.L. Daletskii, Functional integrals connected with operator evolution equations. Russ. Math. Surv. **17**(5), 1–107 (1962)
36. F. Guerra, Structural aspects of stochastic mechanics and stochastic field theory. Phys. Rept. **77**, 263–312 (1981)
37. M. Pavon, A new formulation of stochastic mechanics. Phys. Lett. A **209**, 143–149 (1995)
38. M. Pavon, Hamilton's principle in stochastic mechanics. J. Math. Phys. **36**, 6774 (1995)
39. M. Pavon, Lagrangian dynamics for classical, Brownian and quantum mechanical particles. J. Math. Phys. **37**, 3375 (1996)
40. M. Pavon, Stochastic mechanics and the Feynman integral. J. Math. Phys. **41**, 6060 (2000)
41. F. Kuipers, Analytic continuation of stochastic mechanics. J. Math. Phys. **63**, 4, 042301 (2022)
42. I. Fényes, Eine Wahrscheinlichkeitstheoretische Begründung und Interpretation der Quantenmechanik. Zeitschrift für Physik **132**, 81 (1952)
43. D. Kershaw, Theory of hidden variables. Phys. Rev. **136**(6B), 1850 (1964)
44. E. Madelung, Eine anschauliche Deutung der Gleichung von Schrödinger. Naturwissenschaften **14**, 45, 1004 (1926)
45. E. Madelung, Quantentheorie in hydrodynamischer form. Zeitschrift für Physik **40**(3–4), 322–326 (1927)
46. K. Yasue, Quantum mechanics of nonconservative systems. Ann. Phys. **114**(1–2), 479–496 (1978)
47. K. Yasue, Stochastic quantization: a review. Int. J. Theor. Phys. **18**, 861–913 (1979)
48. K. Yasue, Stochastic calculus of variations. J. Funct. Anal. **41**, 3, 327 (1981)
49. K. Yasue, A path probability representation for wave functions. Lett. Math. Phys. **5**, 93–97 (1981)
50. K. Yasue, Quantum mechanics and stochastic control theory. J. Math. Phys. **22**, 1010–1020 (1981)
51. J.C. Zambrini, K. Yasue, Semi-classical quantum mechanics and stochastic calculus of variations. Ann. Phys. **143**(1), 54–83 (1982)
52. J.C. Zambrini, Stochastic dynamics: a review of stochastic calculus. Int. J. Theor. Phys. **24**, 3, 277 (1985)

53. F. Guerra, P. Ruggiero, New interpretation of the Euclidean-Markov field in the framework of physical Minkowski space-time. Phys. Rev. Lett. **31**, 1022–1025 (1973)
54. F. Guerra, P. Ruggiero, A note on relativistic Markov processes. Lett. Nuovo Cimento **23**, 528 (1978)
55. D. Dohrn, F. Guerra, Compatibility between the Brownian metric and the kinetic metric in Nelson stochastic quantization. Phys. Rev. D **31**, 2521–2524 (1985)
56. R. Marra, M. Serva, Variational principles for a relativistic stochastic mechanics. Ann. Inst. H. Poincare Phys. Theor. **53**, 97–108 (1990)
57. P. Garbaczewski, J.R. Klauder, R. Olkiewicz, The Schrodinger problem, Levy processes and all that noise in relativistic quantum mechanics. Phys. Rev. E **51**, 4114–4131 (1995)
58. L.M. Morato, L. Viola, Markov diffusions in comoving coordinates and stochastic quantization of the free relativistic spinless particle. J. Math. Phys. **36**, 4691–4710 (1995) [erratum: J. Math. Phys. **37**, 4769 (1996)]
59. M. Pavon, On the stochastic mechanics of the free relativistic particle. J. Math. Phys. **42**, 4846 (2001)
60. F. Kuipers, Stochastic quantization of relativistic theories. J. Math. Phys. **62**, 12, 122301 (2021)
61. T.G. Dankel, Mechanics on manifolds and the incorporation of spin into Nelson's stochastic mechanics. Arch. Rational. Mech. Anal. **37**, 192 (1971)
62. D. Dohrn, F. Guerra, Nelson's stochastic mechanics on Riemannian manifolds. Lett. Nuovo Cimento **22**, 4, 121 (1978)
63. D. Dohrn, F. Guerra, Geodesic correction to stochastic parallel displacement of tensors, in *Stochastic Behavior in Classical and Quantum Hamiltonian Systems*. Lecture Notes in Physics, vol. 93 (Springer, 1979), pp. 241–249
64. F. Guerra, L.M. Morato, Quantization of dynamical systems and stochastic control theory. Phys. Rev. D **27**, 1774 (1983)
65. T. Koide, T. Kodama, Novel effect induced by spacetime curvature in quantum hydrodynamics. Phys. Lett. A **383**, 2713–2718 (2019)
66. F. Kuipers, Stochastic quantization on lorentzian manifolds. JHEP **05**, 028 (2021)
67. F. Kuipers, Spacetime stochasticity and second order geometry, in *Lie Theory and Its Applications in Physics. LT XIV, Proceedings in Mathematics & Statistics*, ed. by V. Dobrev, vol. 396 (Springer, Singapore, 2022)
68. P.A. Meyer, A differential geometric formalism for the Itô calculus. *Stochastic Integrals*. Lecture Notes in Mathematics, vol. 851 (Springer, 1981)
69. L. Schwartz, *Semi-Martingales and their Stochastic Calculus on Manifolds* (Presses de l'Université de Montréal, 1984)
70. M. Emery, *Stochastic Calculus in Manifolds* (Springer, 1989)
71. Q. Huang, J.C. Zambrini, *From Second-order Differential Geometry to Stochastic Geometric Mechanics* (2022). arXiv:2201.03706 [math-ph]
72. D. Dohrn, F. Guerra, P. Ruggiero, *Spinning Particles and Relativistic Particles in the Framework of Nelson's Stochastic Mechanics*. Lecture Notes in Physics, vol. 106 (Springer, 1979)
73. W.G. Faris, Spin correlation in stochastic mechanics. Found. Phys. **12**, 1–26 (1982)
74. G.F. De Angelis, G. Jona-Lasinio, M. Serva, N. Zanghi, Stochastic mechanics of a Dirac particle in two space-time dimensions. J. Phys. A **19**, 865–871 (1986)
75. K. Yasue, Stochastic quantization of wave fields and its application to dissipatively interacting fields. J. Math. Phys. **19**, 1892 (1978)
76. F. Guerra, M.I. Loffredo, Stochastic equations for the Maxwell field. Lett. Nuovo Cimento **27**, 41–45 (1980)
77. M.P. Davidson, The generalized Fenyes-Nelson model for free scalar field theory. Lett. Math. Phys. **4**, 101–106 (1980)
78. M. Davidson, Stochastic quantization of the linearized gravitational field. J. Math. Phys. **23**, 132–137 (1982)
79. S. De Siena, P. Ruggiero, F. Guerra, Stochastic quantization of the vector meson field. Phys. Rev. D **27**, 2912–2915 (1983)

80. S. De Siena, F. Guerra, P. Ruggiero, On the connection between the stochastic quantization of the vector meson field and the Euclidean theory. Phys. Rev. D **33**, 2498–2499 (1986)
81. E. Nelson, Stochastic mechanics of relativistic fields. J. Phys. Conf. Ser. **504**, 012013 (2014)
82. T. Kodama, T. Koide, *Variational Principle of Hydrodynamics and Quantization by Stochastic Process* (2014). arXiv:1412.6472 [quant-ph]
83. T. Koide, T. Kodama, K. Tsushima, *Stochastic Variational Method as a Quantization Scheme II: Quantization of Electromagnetic Fields* (2014). arXiv:1406.6295 [hep-th]
84. T. Koide, T. Kodama Stochastic variational method as quantization scheme: field quantization of the complex Klein-Gordon equation. Prog. Theor. Exp. Phys. **2015**, 9, 093A03 (2015)
85. D. de Falco, S. de Martino, S. de Siena, Position-Momentum uncertainty relations in stochastic mechanics. Phys. Rev. Lett. **49**, 181 (1982) [erratum: Phys. Rev. Lett. **50**, 704 (1983)]
86. S. Golin, Uncertainty relations in stochastic mechanics. J. Math. Phys. **26**, 2781 (1985)
87. H.H. Rosenbrock, A stochastic variational principle for quantum mechanics. Phys. Lett. A **110**, 343–346 (1986)
88. H.H. Rosenbrock, The definition of state in the stochastic variational treatment of quantum mechanics. Phys. Lett. A **254**, 307–313 (1999)
89. N.C. Petroni, L.M. Morato, Entangled states in stochastic mechanics. J. Phys. A **33**(33), 5833–5848 (2000)
90. L.S.F. Olavo, L.C. Lapas, A. Figueiredo, Foundations of quantum mechanics: the Langevin equations for QM. Ann. Phys. **327**, 5, 1391 (2012)
91. P. de la Pena, A.M. Cetto, A. Valdes Hernandez, *The Emerging Quantum* (Springer International Publishing, 2015)
92. T. Koide, T. Kodama, Generalization of uncertainty relation for quantum and stochastic systems. Phys. Lett. A **382**, 1472–1480 (2018)
93. J.P. Gazeau, T. Koide, Uncertainty relation for angle from a quantum-hydrodynamical perspective. Ann. Phys. **416**, 168159 (2020)
94. L.F. Santos, C.O. Escobar, Stochastic motion of an open bosonic string. Phys. Lett. A **256**, 89–94 (1999)
95. L. Smolin, Matrix models as nonlocal hidden variables theories. AIP Conf. Proc. **607**(1), 244–261 (2002)
96. F. Markopoulou, L. Smolin, Quantum theory from quantum gravity. Phys. Rev. D **70**, 124029 (2004)
97. J. Erlich, Stochastic emergent quantum gravity. Class. Quant. Grav. **35**, 24, 245005 (2018)
98. J. Erlich, *A First Analysis of Stochastic Composite Gravity* (2022). arXiv:2208.10268 [gr-qc]
99. T. C. Wallstrom, On the derivation of the Schrödinger equation from Stochastic Mechanics. Found. Phys. Lett. **2**, 2, 113 (1988)
100. T.C. Wallstrom, Inequivalence between the Schrödinger equation and the Madelung hydrodynamic equations. Phys. Rev. A **49**, 3, 1613 (1993)
101. I. Schmelzer, An answer to the Wallstrom objection against Nelsonian stochastics (2021). arXiv:1101.5774 [quant-ph]
102. M. Derakhshani, *Stochastic Mechanics Without Ad Hoc Quantization: Theory And Applications To Semiclassical Gravity* (2017). arXiv:1804.01394 [quant-ph]
103. E. Nelson, Field theory and the future of stochastic mechanics, in *"Stochastic Processes in Classical and Quantum Systems*. Lecture Notes in Physics. ed. by S. Albeverio, G. Casati, D. Merlini, vol. 262 (Springer, Berlin, Heidelberg, 1986)
104. E. Nelson, Review of stochastic mechanics. J. Phys.: Conf. Ser. **361**, 012011 (2012)
105. P. Blanchard, S. Golin, M. Serva, On repeated measurements in stochastic mechanics. Phys. Rev. D **34**, 3732 (1986)
106. M. Derakhshani, G. Bacciagaluppi, *On Multi-time Correlations in Stochastic Mechanics* (2022). arXiv:2208.14189 [quant-ph]
107. M. Born, Quantenmechanik der Stoßvorgänge. Z. Phys. **38**(11–12), 803–827 (1926)
108. B.S. DeWitt, Dynamical theory in curved spaces I: a review of the classical and quantum action principles. Rev. Mod. Phys. **29**, 377–397 (1957)

109. W. Pauli, *Pauli Lectures on Physics 6: Selected Topics in Field Quantization* (MIT Press, 1973)
110. J.S. Bell, On the Einstein-Podolsky-Rosen paradox. Phys. Physique Fizika **1**, 195–200 (1964)
111. J.S. Bell, On the problem of hidden variables in quantum mechanics. Rev. Mod. Phys. **38**, 447–452 (1966)
112. B. Schulz, *A New Look at Bell's Inequalities and Nelson's Theorem* (2008). arXiv:0807.3369v30 [quant-ph]
113. F. Calogero, Cosmic origin of quantization. Phys. Lett. A **228**(6), 335–346 (1997)
114. F. Calogero, Cosmic origin of quantization. Int. J. Mod. Phys. B **18**(4–5), 519–525 (2004)
115. B. Mandelbrot, An outline of a purely phenomenological theory of statistical thermodynamics-I: canonical ensembles. IEEE Trans. Inform. Theory **2**, 190–203 (1956)
116. Z. Roupas, Thermodynamic origin of quantum time-energy uncertainty relation. J. Stat. Mech. **2109**, 093207 (2021)
117. M. Arzano, J. Kowalski-Glikman, *Deformations of Spacetime Symmetries: Gravity, Group-Valued Momenta, and Non-commutative Fields*, Lecture Notes in Physics, vol. 986. (Springer, Heidelberg, 2021)
118. I. Karatzas, S.E. Schreve, *Brownian Motion and Stochastic Calculus Graduate Texts in Mathematics* (Springer, New York, 1998)
119. E. Çinlar, *Probability and Stochastics Graduate Texts in Mathematics* (Springer, New York, 2011)
120. S.E. Schreve, *Stochastic Calculus for Finance I: The Binomial Asset Pricing Model* (Springer Finance, Springer, New York, 2004)
121. S.E. Schreve, *Stochastic Calculus for Finance II: Continuous-Time Models* (Springer Finance, Springer, New York, 2004)

Printed in the United States
by Baker & Taylor Publisher Services